Entropy in Dynamic Systems

Entropy in Dynamic Systems

Special Issue Editors

Jan Awrejcewicz
José A. Tenreiro Machado

MDPI • Basel • Beijing • Wuhan • Barcelona • Belgrade

Special Issue Editors

Jan Awrejcewicz
Lodz University of Technology
Poland

José A. Tenreiro Machado
Polytechnic Institute of Porto
Portugal

Editorial Office
MDPI
St. Alban-Anlage 66
4052 Basel, Switzerland

This is a reprint of articles from the Special Issue published online in the open access journal *Entropy* (ISSN 1099-4300) form 2018 to 2019 (available at: https://www.mdpi.com/journal/entropy/special_issues/Entropy_Dynamic_Systems)

For citation purposes, cite each article independently as indicated on the article page online and as indicated below:

LastName, A.A.; LastName, B.B.; LastName, C.C. Article Title. *Journal Name* **Year**, *Article Number*, Page Range.

ISBN 978-3-03921-616-1 (Pbk)
ISBN 978-3-03921-617-8 (PDF)

Contents

About the Special Issue Editors

Jan Awrejcewicz (JA), Professor, holds B.E., M.Sc., Ph.D., and Sc.D. degrees in Mechanical Engineering from the Lodz University of Technology. He is Head of the Department of Automation, Biomechanics and Mechatronics and a corresponding member of the Polish Academy of Sciences. He has won numerous national/international awards including the Humboldt Award (twice). To date, he has supervised 26 doctoral students, published 24 authored/co-authored monographs (in English), over 400 journal papers, and 160 book chapters; he has given also 75 Keynote/plenary talks. JA has received honorary doctorates from the Universities of Bielsko-Biala, Czestochowa, Kielce, Kharkiv, and Gdansk. His research covers mechanics, physics and applied mathematics. For more information, please visit the website www.abm.p.lodz.pl.

José A. Tenreiro Machado, Professor, was born in 1957. He graduated with 'Licenciatura' (1980), Ph.D. (1989) and 'Habilitation' (1995), in Electrical and Computer Engineering from the University of Porto. In 1980–1998, he worked at the Dept. of Electrical and Computer Engineering of the University of Porto. Since 1998, he has worked at the Institute of Engineering, Polytechnic Institute of Porto. He is presently Principal Coordinator Professor at the Dept. of Electrical Engineering, Institute of Engineering, Polytechnic of Porto, Portugal. His research interests include complex systems, nonlinear dynamics, fractional calculus, entropy, modeling, control, data series analysis, biomathematics, and robotics.

Editorial

Entropy in Dynamic Systems

Jan Awrejcewicz [1],* and José A. Tenreiro Machado [2]

1 Department of Automation, Biomechanics and Mechatronics, Lodz University of Technology, 1/15 Stefanowski St., 90-924 Lodz, Poland
2 Department of Electrical Engineering, Institute of Engineering, Polytechnic Institute of Porto, 4249-015 Porto, Portugal
* Correspondence: jan.awrejcewicz@p.lodz.pl

Received: 30 August 2019; Accepted: 12 September 2019; Published: 16 September 2019

In order to measure and quantify the complex behavior of real-world systems, either novel mathematical approaches or modifications of classical ones are required to precisely predict, monitor and control complicated chaotic and stochastic processes. Though the term of entropy comes from Greek and emphasizes its analogy to energy, nowadays, it has wandered to different branches of pure and applied sciences and is understood in a rather rough way with emphasis put on the transition from regular to chaotic states, stochastic and deterministic disorder, uniform and non-uniform distribution or decay of diversity.

This Special Issue originates from the 14th International Conference "Dynamical Systems – Theory and Applications", held December 11–14, 2017 in Łódź (Poland), and addresses the notion of entropy in a very broad sense. The presented manuscripts follow from different branches of mathematical/physical sciences, natural/social sciences and engineering-oriented sciences with emphasis put on the complexity of dynamical systems. Topics like timing chaos and spatiotemporal chaos, bifurcation, synchronization and anti-synchronization, stability, lumped mass and continuous mechanical systems modeling, novel non-linear phenomena, and resonances are discussed.

An Analysis of Deterministic Chaos as an Entropy Source for Random Number Generators by Kaya Demir and Salih Ergün [1] presents the results of a comparison between bounded chaos, unbounded chaos and Gaussian white noise as a source of entropy for a random number of generators yielded by an analytical study of the autocorrelation and the approximate entropy analysis of the resulting bit sequences [2].

Information Transfer Among the Components in Multi-Dimensional Complex Dynamical Systems by Yimin Yin and Xiaojun Duan [3] provides a rigorous formalism of information transfer within a multi-dimensional deterministic dynamic system established for both continuous flows and discrete mappings.

Fractional Form of a Chaotic Map without Fixed Points: Chaos, Entropy and Control by Adel Ouannas, Xiong Wang, Amina-Aicha Khennaoui, Samir Bendoukha, Viet-Thanh Pham and Fawaz E. Alsaadi [4] presents the results of the first dynamic investigation of a fractional order chaotic map corresponding to a recently developed standard map that exhibits chaotic behavior with no fixed point. The authors use the approximate entropy measure to quantify the level of chaos in the fractional map.

Tsallis Entropy of Product MV-Algebra Dynamical Systems by Dagmar Markechová and Beloslav Riečan [5] provides an example of the mathematical modelling of Tsallis of product MV-algebra dynamical entropy to provide the entropy measure that is invariant under isomorphism.

A Novel Image Encryption Scheme Based on Self-Synchronous Chaotic Stream Cipher and Wavelet Transform by Chunlei Fan and Qun Ding [6] presents a self-synchronous chaotic stream cipher that ensures the limited error propagation of image data in the secure transmission of image data. The cipher is designed with the purpose of resisting active attack.

The General Solution of Singular Fractional-Order Linear Time-Invariant Continuous Systems with Regular Pencils by Iqbal M. Batiha, Reyad El-Khazali, Ahmed AlSaedi and Shaher Momani [7] proposes the use of the Adomian decomposition method based on the Caputo's definition of the fractional-order derivative to obtain a general solution of singular fractional-order linear-time invariant continuous systems.

Quantifying Chaos by Various Computational Methods. Part 1: Simple Systems by Jan Awrejcewicz, Anton V. Krysko, Nikolay P. Erofeev, Vitalyj Dobriyan, Marina A. Barulina and Vadim A. Krysko [8] proposes an algorithm of calculation of the spectrum of Lyapunov exponents based on a trained neural network that can be used to compute a spectrum of Lyapunov exponents, and then to detect a transition of the system regular dynamics into chaos, hyperchaos, and others.

Quantifying Chaos by Various Computational Methods. Part 2: Vibrations of the Bernoulli–Euler Beam Subjected to Periodic and Colored Noise by Jan Awrejcewicz, Anton V. Krysko, Nikolay P. Erofeev, Vitalyi Dobriyan, Marina A. Barulina and Vadim A. Krysko [9] presents a theory of non-linear dynamics of flexible Euler–Bernoulli beams under transverse harmonic load and colored noise that has been extended to investigate a novel transition type exhibited by non-equilibrium systems embedded in a stochastic fluctuated medium.

On Points Focusing Entropy by Ewa Korczak-Kubiak, Anna Loranty and Ryszard J. Pawlak [10] introduces the notion of a (asymptotical) focal entropy point allowing study of the local aspects of the entropy of non-autonomous dynamical systems.

Last but not least, Analytical Solutions for Multi-Time Scale Fractional Stochastic Differential Equations Driven by Fractional Brownian Motion and Their Applications by Xiao-Li Ding and Juan J. Nieto [11] presents a method of investigation of the analytical solutions of multi-time scale fractional stochastic differential equations driven by fractional Brownian motions that is continued in [12].

Conflicts of Interest: The authors declare no conflict of interest.

References

1. Demir, K.; Ergün, S. An Analysis of Deterministic Chaos as an Entropy Source for Random Number Generators. *Entropy* **2018**, *20*, 957. [CrossRef]
2. Butusov, D.; Karimov, A.; Tutueva, A.; Kaplun, D.; Nepomuceno, E.G. The effects of Padé numerical integration in simulation of conservative chaotic systems. *Entropy* **2019**, *21*, 362. [CrossRef]
3. Yin, Y.; Duan, X. Information Transfer Among the Components in Multi-Dimensional Complex Dynamical Systems. *Entropy* **2018**, *20*, 774. [CrossRef]
4. Ouannas, A.; Wang, X.; Khennaoui, A.; Bendoukha, S.; Pham, V.; Alsaadi, F. Fractional Form of a Chaotic Map without Fixed Points: Chaos, Entropy and Control. *Entropy* **2018**, *20*, 720. [CrossRef]
5. Markechová, D.; Riečan, B. Tsallis Entropy of Product MV-Algebra Dynamical Systems. *Entropy* **2018**, *20*, 589. [CrossRef]
6. Fan, C.; Ding, Q. A Novel Image Encryption Scheme Based on Self-Synchronous Chaotic Stream Cipher and Wavelet Transform. *Entropy* **2018**, *20*, 445. [CrossRef]
7. Batiha, I.; El-Khazali, R.; AlSaedi, A.; Momani, S. The General Solution of Singular Fractional-Order Linear Time-Invariant Continuous Systems with Regular Pencils. *Entropy* **2018**, *20*, 400. [CrossRef]
8. Awrejcewicz, J.; Krysko, A.; Erofeev, N.; Dobriyan, V.; Barulina, M.; Krysko, V. Quantifying Chaos by Various Computational Methods. Part 1: Simple Systems. *Entropy* **2018**, *20*, 175. [CrossRef]
9. Awrejcewicz, J.; Krysko, A.; Erofeev, N.; Dobriyan, V.; Barulina, M.; Krysko, V. Quantifying Chaos by Various Computational Methods. Part 2: Vibrations of the Bernoulli–Euler Beam Subjected to Periodic and Colored Noise. *Entropy* **2018**, *20*, 170. [CrossRef]
10. Korczak-Kubiak, E.; Loranty, A.; Pawlak, R. On Points Focusing Entropy. *Entropy* **2018**, *20*, 128. [CrossRef]

11. Ding, X.; Nieto, J. Analytical Solutions for Multi-Time Scale Fractional Stochastic Differential Equations Driven by Fractional Brownian Motion and Their Applications. *Entropy* **2018**, *20*, 63. [CrossRef]
12. Ding, X.-L.; Nieto, J.J. Analysis and Numerical Solutions for Fractional Stochastic Evolution Equations with Almost Sectorial Operators. *J. Comput. Nonlinear Dynam.* **2019**, *14*, 091001. [CrossRef]

Article

An Analysis of Deterministic Chaos as an Entropy Source for Random Number Generators

Kaya Demir and Salih Ergün *

TÜBİTAK—Informatics and Information Security Research Center, 41470 Gebze, Kocaeli, Turkey; kaya.demir@tubitak.gov.tr
* Correspondence: salih.ergun@tubitak.gov.tr; Tel.: +90-262-675-1370

Received: 6 October 2018; Accepted: 25 October 2018; Published: 11 December 2018

Abstract: This paper presents an analytical study on the use of deterministic chaos as an entropy source for the generation of random numbers. The chaotic signal generated by a phase-locked loop (PLL) device is investigated using numerical simulations. Depending on the system parameters, the chaos originating from the PLL device can be either bounded or unbounded in the phase direction. Bounded and unbounded chaos differs in terms of the flatness of the power spectrum associated with the chaotic signal. Random bits are generated by regular sampling of the signal from bounded and unbounded chaos. A white Gaussian noise source is also sampled regularly to generate random bits. By varying the sampling frequency, and based on the autocorrelation and the approximate entropy analysis of the resulting bit sequences, a comparison is made between bounded chaos, unbounded chaos and Gaussian white noise as an entropy source for random number generators.

Keywords: deterministic chaos; random number generator; unbounded chaos; bounded chaos; phase-locked loop; Gaussian white noise

---

1. Introduction

Random number generators (RNGs) are fundamental components of cryptographic systems, as they are responsible for generating the unpredictable key values used in ciphering algorithms to protect the integrity, confidentiality and authenticity of the information [1]. Basically, an RNG consists of an entropy source, a sampler to harvest entropy and a post processor to remove statistical imperfections [2]. An ideal entropy source used in an RNG system should have a constant power spectral density over its operating bandwidth, and it is preferable that this bandwidth is as wide as possible [3]. A commonly used entropy source in RNGs is based on amplification of a physical noise in the microscopic domain, such as thermal or shot noise [4,5]. However, it has been previously demonstrated that chaotic noise obtained from a macroscopic system can also be used to generate white noise, eliminating the need for amplification [6]. Despite being deterministic, chaotic systems can be used as an entropy source due to their extreme sensitivity to initial conditions, noise-like power spectrum and positive Lyapunov exponent [7]. The use of chaotic systems as an entropy source in RNGs suggests the possibility of reaching higher throughput data without the need for post processing and with ease of implementation in an integrated circuit form [8–10]. With two or more positive Lyapunov exponents, hyperchaotic systems can also be used for random number generation, and they have more complex behaviors, making it harder to predict the RNG output time series [11]. However, synchronization of two coupled hyperchaotic systems despite parameter mismatches was demonstrated in [12] based on the concept of the Master Stability Function. In [13], the security issues of chaos-based random number generators were discussed by studying the synchronization of chaotic systems, and it is suggested that the inclusion of noise analysis in deterministic chaos qualifies chaos-based generators as a truly random number source. In this paper, the use of a phase-locked loop (PLL) device in the chaotic regime for random number generation is considered. The use of PLL circuits for the generation of random numbers in reconfigurable hardware platforms was extensively studied, where the randomness was extracted from the intrinsic jitter of the synthesized clock signal by the PLL [14–17]. In this paper, the PLL device is used to generate bounded and unbounded chaos, as described in [18,19]. At the input of the voltage-controlled oscillator (VCO) component of the PLL, a chaotic signal is observed under certain conditions; however, the spectrum of this chaotic signal is non-flat over the spectral bandwidth. By adjusting the system parameters, the chaos can be made unbounded, which results in an approximately flat spectrum up to a certain frequency, similar to white Gaussian noise [20].

In this paper, using deterministic chaos as an entropy source, random bit sequences of fixed length are generated by regularly sampling the chaotic signal observed at the VCO input when the nature of chaos is bounded and unbounded. The sampling frequency limits the maximum throughput of the RNG and is a critical parameter to ensure randomness in the generated bit stream. A faster sampling rate is preferred for high-throughput data, but the bandwidth of the entropy source imposes restrictions on the maximum allowable sampling frequency to maintain randomness in the resulting bit sequence. To investigate this phenomenon, the sampling frequency is gradually elevated, and the randomness of the resulting bit streams are assessed through the application of the concepts of autocorrelation and approximate entropy. Then the bounded and unbounded chaos is benchmarked against white Gaussian noise, which might originate in a stationary stochastic process. It is numerically shown that up to a certain frequency which is dependent on the PLL system parameters, unbounded chaos approaches white Gaussian noise as an entropy source to generate random numbers by the regular sampling of an irregular waveform method. To the best of authors' knowledge, this is the first analytical study on the application of bounded and unbounded chaos from a PLL device as an entropy source for RNG and the comparison of deterministic bounded or unbounded chaos and white Gaussian noise from a stochastic process. This paper is organized as follows. In Section 2, the equations governing the PLL system are described. Section 3 focuses on the formation of bounded and unbounded chaos in PLL and the analysis of the associated chaotic signals. In Section 4, regular sampling of the chaotic signal method is applied to generate random bit sequences using bounded chaos, unbounded chaos and white Gaussian noise as an entropy source and the randomness of the generated bit sequences are discussed, followed by conclusions in Section 5.

2. Chaotic System

The chaotic system in this paper is based on a sinusoidally driven PLL, which has previously been extensively studied in [19,20]. Therefore, the equations will be summarized by referring to [19,20]. Basically, a PLL device is made up of a phase detector to identify the phase error, a low-pass filter and a VCO generating a square wave at a frequency dependent on input signal amplitude. Figure 1 illustrates the phase model of a PLL. The θ_{in}, θ_{out} and $\phi(t) = \theta_{in} - \theta_{out}$ are input phase, output phase and phase error, respectively.

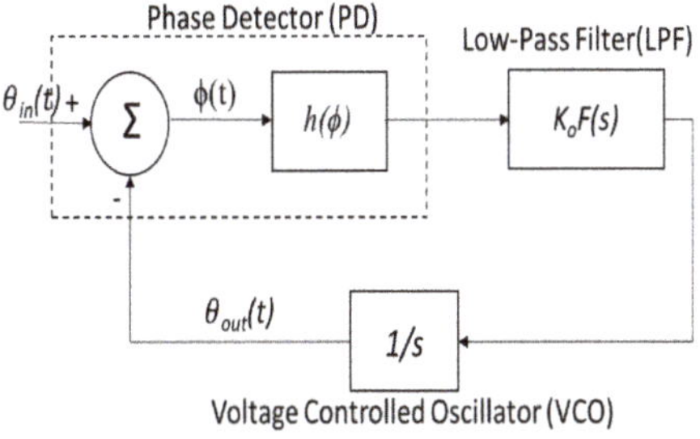

Figure 1. The phase model of the PLL system.

The nonlinear function $h(\phi)$ is a 2π-periodic function of ϕ and it is in symmetric triangular form for an EX-OR type phase detector. It is defined as

$$h(\phi) = h(\phi + 2n\pi), \quad n = 0, \pm1, \pm2, \ldots .n \in \mathbb{Z} \tag{1}$$

$$h(\phi) = \begin{cases} \phi & for \ |\phi| < \frac{\pi}{2} \\ -\phi + \pi \ for \ \frac{\pi}{2} < \phi < \frac{3\pi}{2} \end{cases} \tag{2}$$

The transfer function of the low-pass (lag-lead) filter is given as

$$F(s) = (1 + \tau_2 s)/(1 + \tau_1 s) \tag{3}$$

Following the diagram in s-domain

$$\phi(s) = \theta_{in}(s) - \theta_{out}(s) \tag{4}$$

$$\theta_{out}(s) = h(\phi(s))K_0 F(s) \cdot \frac{1}{s} \tag{5}$$

Substituting (3), (5) in (4), using cross multiplication and then dividing each side by τ_1,

$$\phi(s) + h(\phi(s))K_0 \frac{(1 + \tau_2 s)}{(1 + \tau_1 s)s} = \theta_{in}(s) \tag{6}$$

$$\tau_1 s^2 \phi(s) + s\phi(s) + K_0 h(\phi(s)) + \tau_2 s K_0 h(\phi(s)) = \tau_1 s^2 \theta_{in}(s) + s\theta_{in}(s) \tag{7}$$

$$s^2 \phi(s) + \frac{1}{\tau_1}(s\phi(s) + K_0 \tau_2 sh(\phi(s))) + \left(\frac{K_0}{\tau_1}\right)h(\phi(s)) = s^2 \theta_{in}(s) + \frac{1}{\tau_1}s\theta_{in}(s) \tag{8}$$

Using (8), with respect to the phase error ϕ, the system equation in the time domain can be stated as

$$\frac{d^2\phi}{dt^2} + \frac{1}{\tau_1}(1 + K_0 \tau_2 h'(\phi))\frac{d\phi}{dt} + \left(\frac{K_0}{\tau_1}\right)h(\phi) = \frac{d^2\theta_{in}}{dt^2} + \frac{1}{\tau_1}\frac{d\theta_{in}}{dt} \tag{9}$$

Assuming the input signal is modulated by a sinusoidal waveform

$$\frac{d\theta_{in}}{dt} = \Delta\omega + M sin\omega_m t \tag{10}$$

$$\Delta\omega = \omega_{in} - \omega_{out} \tag{11}$$

where $\omega_m, \omega_{in}, \omega_{out}$ and $\Delta\omega$ are phase modulation, input signal, output signal angular frequencies and frequency detuning, respectively. The natural angular frequency and the damping coefficient are defined as follows:

$$\omega_n = \sqrt{K_0/\tau_1} = 2\pi f_n \tag{12}$$

$$\zeta = (1 + K_0 \tau_2)/2\sqrt{K_0 \tau_1} \tag{13}$$

To simplify the equations, the following normalized parameters are introduced:

$$\beta = \frac{\omega_n}{K_0} = \frac{1}{\sqrt{K_0 \tau_1}} \quad \text{normalized natural frequency} \tag{14a}$$

$$\sigma = \frac{\Delta\omega}{\omega_n} \quad \text{normalized frequency detuning} \tag{14b}$$

$$\Omega_m = \frac{\omega_m}{\omega_n} \quad \text{normalized modulation frequency} \tag{14c}$$

$$m = \frac{M}{\omega_n} \quad \text{normalized maximum angular frequency deviation} \tag{14d}$$

By changing the time t into $\tau = \omega_n t$ and replacing τ by t again, Equation (9) can be given as

$$\frac{d^2\phi}{dt^2} + \beta\left[1 + \frac{(2\zeta - \beta)h'(\phi)}{\beta}\right]\frac{d\phi}{dt} + h(\phi) = \beta\sigma + \beta\,m sin\Omega_m t + m\Omega_m cos\Omega_m t \tag{15}$$

where $2\zeta - \beta = K_0\tau_2/\sqrt{K_0\tau_1} \geq 0$. For simplicity, the filter is assumed to be a lag filter $2\zeta - \beta = 0, (\tau_2 = 0)$. By changing the time t into $t = t' - \frac{\theta}{\Omega_m}$ where $\tan(\theta) = \frac{\Omega_m}{\beta}$, and replacing t' by t again, Equation (15) can be simplified as

$$\frac{d^2\phi}{dt^2} + \beta\frac{d\phi}{dt} + h(\phi) = \beta\sigma + a sin\Omega_m t \tag{16}$$

$$a = m\sqrt{\beta^2 + \Omega_m^2} \tag{17}$$

For small values of a, the solutions are periodic with Ω_m, which means that the phase ϕ is phase-locked with the input signal. With an increase in a, the phase ϕ shows bifurcations and becomes chaotic, as seen in Figure 2. For fixed values of β and Ω_m, the parameter m is gradually increased, thus the a parameter is linearly elevated. The bifurcations and transition from bounded to unbounded chaos is observed in Figure 2. It is observed that the PLL system demonstrates unbounded chaos when m (normalized maximum angular frequency deviation) is between approximately 1.75 and 3.

Depending on the parameters as seen in Figure 2, the chaos is either bounded or unbounded in the ϕ-direction. In [6,18–20], it is suggested that the chaotic change of $\phi(t)$ approaches a Wiener-Levy process over long times. Therefore, its derivative $d\phi/dt$ is supposed to yield a white noise-like spectrum at angular frequencies substantially below Ω_m, $\omega_0(\omega_0 = 1)$ and $\omega_r(\omega_r = \beta/2)$, where ω_0 and ω_r are the natural angular frequency and the relaxation angular frequency of the simplified system, respectively. The chaotic signal at the VCO input is given as

$$\dot{\theta}_{out} = m sin\Omega_m t - \dot{\phi} \tag{18}$$

In this paper, the focus is on the signal at the VCO input as it exhibits chaotic behaviors which can be exploited for random number generation.

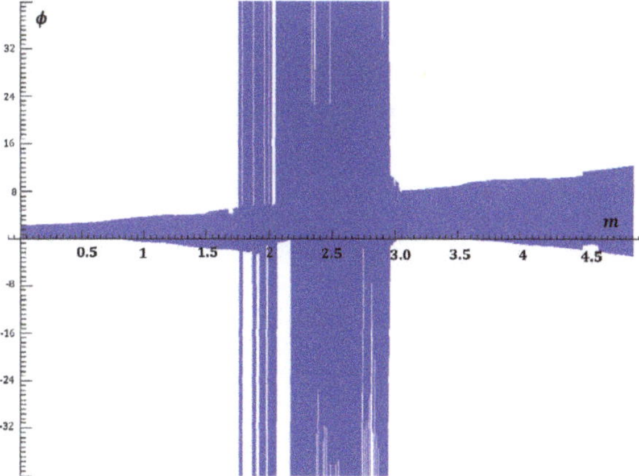

Figure 2. Bifurcation map of ϕ with respect to m illustrating bounded and unbounded chaos ($\beta = 0.56$, $\Omega_m = 0.9$, $\sigma = 0$).

3. Chaotic Signal Formation

In this section, the normalized equations of the PLL are entered into a numerical solver, and the system parameters are adjusted to set the PLL to operate in the chaotic regime. For numerical solutions, Dynamics Solver software is used. To put the system in chaos, the normalized natural frequency β and the normalized modulation frequency Ω_m are chosen to be 0.56 and 0.9, respectively, although many other combinations of parameter values for chaos can be found by experimenting with the numerical solver. Then, a is changed linearly by varying the m parameter according to (17). The bifurcation graph shown in Figure 2 suggests that for these values of β and Ω_m, $m = 1.75$ is the transition border from bounded to unbounded chaos. Figure 3 illustrates the change of Lissajous patterns in ϕ-$\dot{\phi}$ plane when the type of chaos transitions from bounded at $m = 1.74$ to unbounded at $m = 2.29$ by modifying only the m parameter in parallel with the bifurcation graph shown in Figure 2.

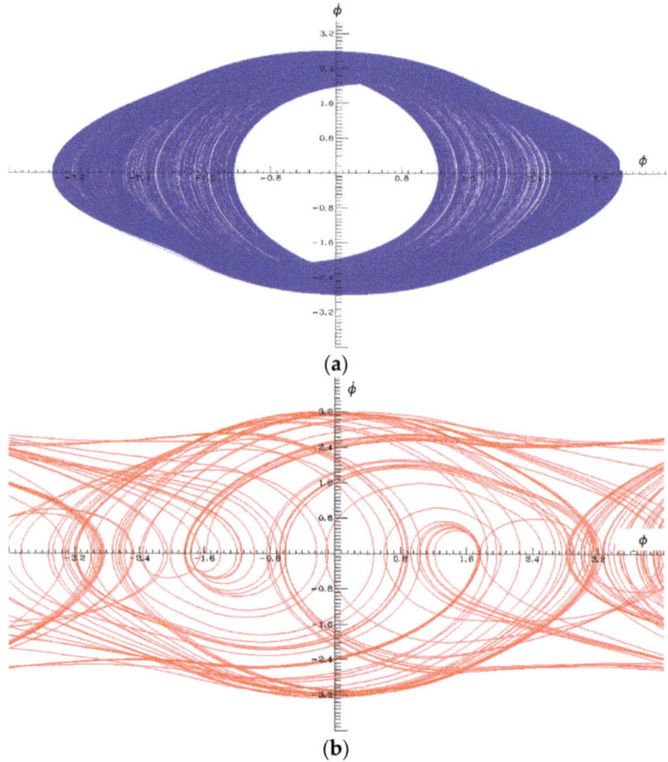

Figure 3. Lissajous patterns in $\phi - \dot{\phi}$ plane exhibiting (**a**) bounded and (**b**) unbounded chaos. Associated parameters $\beta = 0.56$, $\Omega_m = 0.9$, $\sigma = 0$ and $m = (1.74, 2.29)$ for bounded and unbounded chaos respectively.

Figures 4 and 5 show the signals at VCO input in the time domain and the associated power spectra for cases of unbounded and bounded chaos in a comparative manner. As the solutions are obtained for normalized equations, units are not shown in Figure 4 and 5. Even by observing the time-domain waveforms, it may be possible to distinguish between unbounded and bounded chaos, as unbounded chaos displays a larger degree of irregularity and aperiodicity compared to bounded chaos. The time-domain signal at the VCO input, as shown in Figure 4, seems visually more regular compared to the signal shown in Figure 5. However, power spectral analysis is a safer method to identify the type of chaos. Figure 4 demonstrates that the power spectrum is not flat for bounded chaos; therefore, it is not an optimal entropy source for random number generation. However, for unbounded chaos, Figure 5 shows that the power spectrum for the VCO input in a normalized equation can be visually considered to be flat from DC up to approximately $\Omega \approx 0.15$, which corresponds to a frequency of $f \approx 0.15 \times f_n \approx 230$ Hz. To expand the flat spectrum range, it is necessary to increase the natural frequency for a fixed Ω_m. Having a flat spectrum similar to white noise makes the VCO input signal from an unbounded chaos case a possible entropy source for the generation of random numbers using the regular sampling of an irregular waveform method. It exhibits very irregular and aperiodic behaviors, which make it unpredictable and useful to be exploited for random number generation.

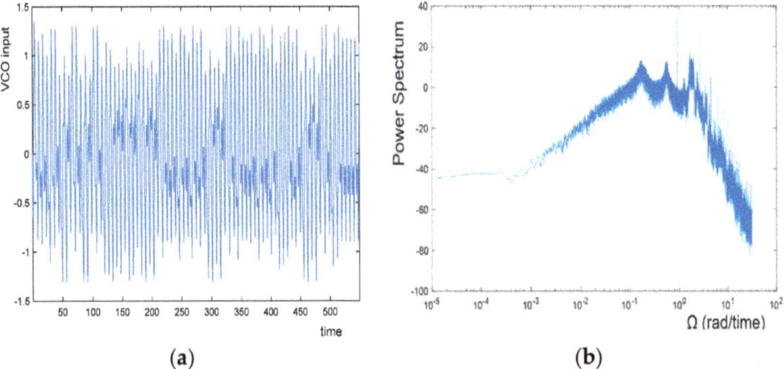

Figure 4. Chaotic signal at VCO input for bounded chaos (**a**) in the time domain, and (**b**) the associated power spectrum.

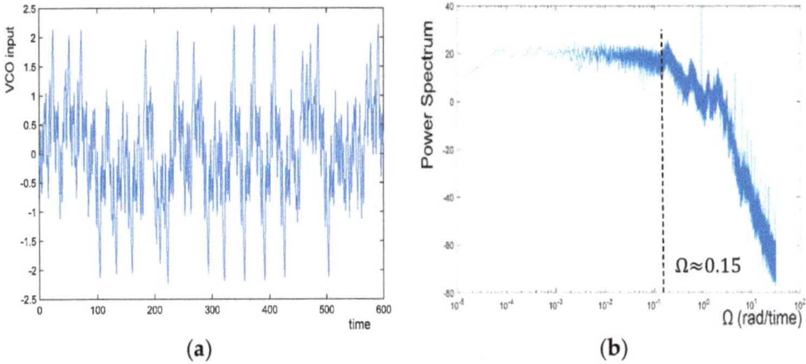

Figure 5. Chaotic signal at VCO input for unbounded chaos (**a**) in the time domain, and (**b**) the associated power spectrum.

4. Random Bit Generation and Discussion

A method to generate output bit sequence is regular sampling of the chaotic signal at the VCO input and comparing the samples with a threshold value. The ergodicity of chaotic signals makes it possible to analyze the distribution and statistical properties of the chaotic variable independent of initial conditions and sampling frequency. Therefore, the mean value of the samples over a long operation time can be selected as the threshold. At the time of sampling, if the signal value is below the threshold, bit *0* is produced; otherwise, the output is bit *1*. However, the sampling frequency f_s needs to be adjusted to generate random bits with high entropy and without correlation between successive bits. The sampling frequency determines the throughput of the RNG, and it heavily depends on the power spectrum of the chaotic signal at the VCO input. For selection of the sampling frequency, autocorrelation analysis can be utilized as a metric to quantify randomness in a finite bit sequence. The absolute value of normalized autocorrelation function at a lag of one sampling period gives the correlation between successive bits. Therefore, bit sequences with a length of 20 KBits are generated by regularly sampling the chaotic signal at varying frequencies and the absolute value of the normalized autocorrelation at one-bit lag, i.e., one sampling period is calculated. To find the absolute value of the normalized autocorrelation at one-bit lag, first the mean of the 20 KBits sequence is subtracted from each binary value in the bit stream and x_n is obtained. Then the autocorrelation function for the resulting bit sequence x_n is normalized such that at zero lag, the value is 1, and then its absolute value is calculated. As the autocorrelation function for the 20 KBits sequence is real and symmetric around zero lag, the value corresponding to one-bit lag can be found by substituting $m = \pm 1$ and taking the absolute value of the result in (20):

$$\hat{R}_{xx}(0) = \sum_{n=0}^{N-m-1} x_n^2 \tag{19}$$

$$\hat{R}_{xx,normalized}(m) = \frac{1}{\hat{R}_{xx}(0)} \begin{cases} \displaystyle\sum_{n=0}^{N-m-1} x_{n+m}\, x_n \; , \; m \geq 0 \\ \hat{R}_{xx}(-m) \qquad , \; m < 0 \end{cases} \tag{20}$$

The VCO input signals from the previous section depicting bounded and unbounded chaos are used for random bit generation by the regular sampling of an irregular waveform method. White Gaussian noise generated by Matlab is also used to generate random bits by regular sampling as a benchmark to assess the use of chaos as an entropy source for RNGs.

Figure 5 shows the relation between the sampling period and the absolute values of the normalized autocorrelation function at one-bit lag according to (20) for the bit sequences generated using bounded chaos, unbounded chaos and Gaussian white noise, respectively. As can be expected, the correlation between successive bits generally decreases when the sampling period increases, which also means reducing the sampling frequency. However, the plot of absolute values of the normalized autocorrelation exhibits local maxima and minima for both bounded and unbounded chaos. It is noteworthy to mention that the variation in the autocorrelation value is higher in the case of bounded chaos than for unbounded chaos. In case of bounded chaos, the peaks are separated from each other by approximately $3.6T_n$, where $T_n = \frac{2\pi}{\omega_n}$. Therefore, this can be interpreted as an indication that it would be easier to make a random generator by sampling the chaotic signal in unbounded chaos. For unbounded chaos, from Figure 5, it can be stated approximately that for obtaining a random bit stream, the sampling period should be adjusted as $T_s = k\cdot 5.\, 2\cdot T_n$, $k = 2, 3, ...$ Furthermore, the performance of unbounded chaos as an entropy source approaches that of white Gaussian noise. As white Gaussian noise is assumed to be white with infinite bandwidth, there is no restriction on the minimum sampling period that can be used to generate a random bit sequence. By proper selection of the sampling frequency of the chaotic signal, the absolute value of normalized autocorrelation values at one-bit lag of the bit sequence can be made close to that of the bit sequence generated by regular sampling of Matlab-based white Gaussian noise which has an infinite and flat power spectrum.

However, as can be observed from Figure 6, in general, the absolute value of autocorrelation of bit sequences at one-bit lag generated by white Gaussian noise are lower compared to bit sequences obtained from unbounded chaos. This is because the sampled chaotic signal is obtained purely by the solution of deterministic equations. However, if the system were implemented experimentally, the non-deterministic thermal and shot noise in electronic components would affect the chaotic trajectories continuously, making the bit stream generated by regular sampling of the chaotic signal non-deterministic, thus resulting in autocorrelation values closer to those of white Gaussian noise [9].

To further analyze the randomness of the RNG output, the concept of approximate entropy (ApEn) is used as a measure of sequential irregularity. For a finite length of bit sets, approximate entropy gives an idea of the randomness, with higher ApEn values indicating a higher level of randomness. ApEn approaches the theoretical maximum information entropy of $ln(2) \approx 0.69$ for a perfectly random bit sequence. The approximate entropy (ApEn) values of order 8 are calculated for the bit sequences of a length of 20 KBits generated by sampling unbounded chaos, bounded chaos and a Gaussian white noise signal at varying sampling frequencies. Figure 7 shows the relation between approximate entropy of the output bit stream and the sampling period for unbounded chaos, bounded chaos and Gaussian white noise, respectively. It is noteworthy to mention that at peak points of high autocorrelation, the approximate entropy is at a local minimum. Furthermore, increasing the sampling period, i.e., slowing the RNG throughput, generally results in higher ApEn value. In the case of bounded chaos, there exists a periodic pattern of ups and downs which is due to the non-flat power spectrum. However, in case of unbounded chaos, the ApEn value reaches the theoretical maximum quickly after a certain sampling rate, does not deviate very much, and approximates the value of white Gaussian noise. Unbounded chaos approaches white Gaussian noise, which shows the advantage of using unbounded chaos as an entropy source instead of bounded chaos.

Table 1. Results of the FIPS-140-2 test suite for RNG based on regular sampling of chaotic signal.

Statistical Tests	p-Value
Frequency	0.777297
Block Frequency	0.543739
Runs	0.041646
Longest Run	0.496469

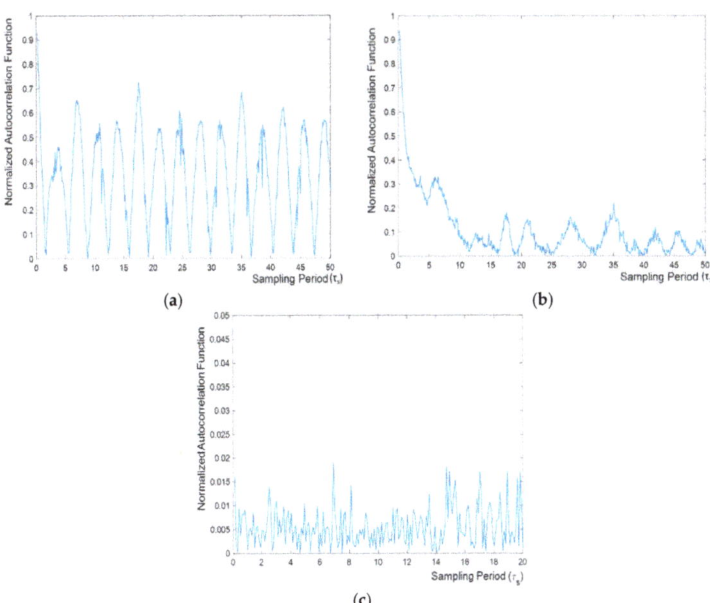

Figure 6. Relation between absolute value of normalized autocorrelation at one-bit lag and the sampling period τ_s for (**a**) bounded chaos, (**b**) unbounded chaos, and (**c**) white Gaussian noise.

As the amount of data that can be numerically produced with regular sampling of a chaotic PLL signal is limited; the data sequence is not put through NIST-800-22 test suite, since it involves at least 40 bit sequences with a length of 1Mbit. Instead, bit streams of 20 KBits length are subjected to tests of FIPS-140-2 test suite without any post processing [7]. The numerical binary streams of length 20 KBits obtained by regular sampling of the chaotic signal from a PLL device passed all 4 statistical tests, as shown in Table 1 where p-Value $(0 \leq p - Value \leq 1)$ is a real number estimating the probability that a perfect RNG would produce a less random sequence than the given sequence.

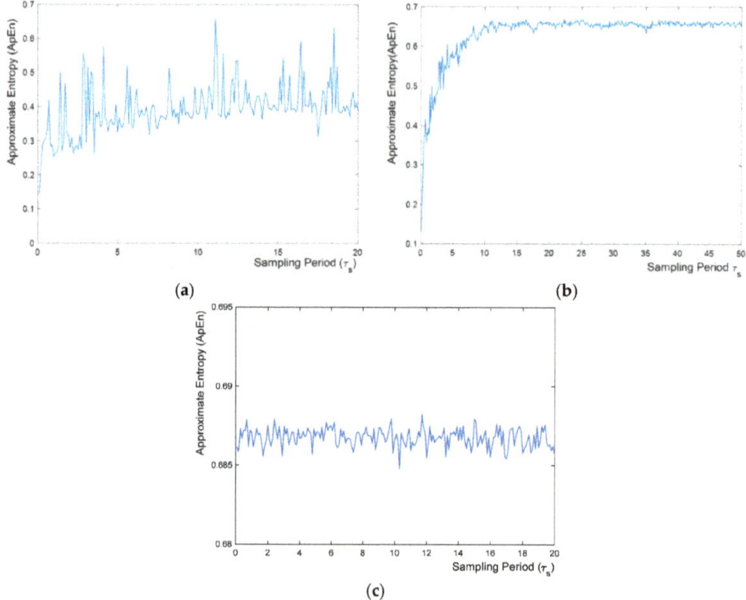

Figure 7. Relation between absolute value of normalized autocorrelation at one-bit lag and the sampling period τ_s for (**a**) bounded chaos, (**b**) unbounded chaos, and (**c**) white Gaussian noise.

5. Conclusions

In this paper, the chaotic signal from a phase-locked loop device is studied in order to develop a comparison between deterministic chaotic and stochastic processes for generation of random numbers. By adjusting the system parameters, the chaos becomes either bounded or unbounded in the phase direction, and this changes the power spectral characteristics of the chaotic signal. Random binary sequences are generated by regular sampling of the chaotic PLL signals. Autocorrelation and approximate entropy analysis is used to quantify the relationship between the randomness of the generated bit streams and the associated sampling periods. The chaotic signals are benchmarked against white Gaussian noise, and it is shown that unbounded chaos approaches Gaussian white noise as an entropy source for random number generation. To the best of the authors' knowledge, this is the first analytical study investigating using bounded and unbounded chaos from a PLL device as an entropy source for random number generation and the comparison of deterministic chaos to white noise from a stochastic process.

Author Contributions: Conceptualization, S.E.; methodology, K.D, S.E.; writing—original draft preparation, K.D.; writing—review and editing, S.E.; supervision, S.E.

Funding: This research received no external funding.

Acknowledgments: This work has been supported by The Scientific and Technological Research Council of Turkey (TÜBİTAK).

Conflicts of Interest: The authors declare no conflict of interest.

References

1. Jun, B.; Kocher, P. *The Intel Random Number Generator*; White Paper; Cryptography Research Inc.: San Francisco, CA, USA, 1999.
2. Menezes, A.; Oorscot, P.V.; Vanstone, S.A. *Handbook of Applied Cryptography*, 1st ed.; CRC Press: Boca Raton, FL, USA, 1996; ISBN 9781439821916.
3. Gov, N.C.; Mihcak, M.K.; Ergün, S. True Random Number Generation via Sampling from Flat Band-Limited Gaussian Process. *IEEE Trans. Circuits Syst. I* **2011**, *58*, 1044–1051. [CrossRef]
4. Holman, W.T.; Connelly, J.A.; Dowlatabadi, A.B. An integrated analog/digital random noise source. *IEEE Trans. Circuits Syst. I* **1997**, *44*, 521–528. [CrossRef]
5. Bagini, V.; Bucci, M. A design of reliable true random number generator for cryptographic applications. In Proceedings of the International Workshop on Cryptographic Hardware and Embedded Systems (CHES), Worcester, MA, USA, 12–13 August 1999; Springer: Berlin, Heidelberg, 1999; pp. 204–218.
6. Kautz, R.L. Using Chaos to Generate White Noise. *J. Appl. Phys.* **1999**, *86*, 5794–5800. [CrossRef]
7. Ergün, S.; Asada, K. Numerical verification of chaos-based random number generators by using bootstrap method. *Nonlinear Theory Its Appl.* **2011**, *2*, 43–53. [CrossRef]
8. Ozoguz, S.; Elwakil, A.S.; Ergün, S. Cross-coupled chaotic oscillators and application to random bit generation. *IEE Proc. Circuits Devices Syst.* **2006**, *153*, 506–510. [CrossRef]
9. Ergün, S.; Güler, Ü.; Asada, K. IC truly random number generators based on regular & chaotic sampling of chaotic waveforms. *IEICE Nonlinear Theory Its Appl.* **2011**, *2*, 246–261.
10. Ergün, S. Modeling and analysis of chaos-modulated dual oscillator-based random number generators. In Proceedings of the European Signal Processing Conference (EUSIPCO), Lausanne, Switzerland, 25–29 August 2008; pp. 1–5.
11. Liu, Y.; Tong, X. Hyperchaotic system-based pseudorandom number generator. *IET Inf. Secur.* **2016**, *10*, 433–441. [CrossRef]
12. Buscarino, A.; Fortuna, L.; Frasca, M. Experimental robust synchronization of hyperchaotic circuits. *Phys. D Nonlinear Phenom.* **2009**, *238*, 1917–1922. [CrossRef]
13. Ergün, S. On the security of chaos based true random generators. *IEICE Trans. Fundam. Electron. Commun. Comput. Sci.* **2016**, *99*, 363–369. [CrossRef]
14. Fischer, V.; Drutarovskỳ, M. True random number generator embedded in reconfigurable hardware. In Proceedings of the International Workshop on Cryptographic Hardware and Embedded Systems (CHES), Redwood Shores, CA, USA, 13–15 August 2002; Springer: Berlin, Heidelberg, 2003; pp. 415–430.
15. Šimka, M.; Drutarovskỳ, M.; Fischer, V. Embedded true random number generator in actel FPGAs. In Proceedings of the Workshop on Cryptographic Advances in Secure Hardware (CRASH), Leuven, Belgium, 6–7 September 2005; pp. 6–7.
16. Varchola, M.; Drutarovskỳ, M.; Fouquet, R.; Fischer, V. Hardware platform for testing performance of TRNGs embedded in actel fusion FPGA. In Proceedings of the 18th International Conference Radioelektronika, Prague, Czech Republic, 24–25 April 2008; pp. 1–4.
17. Bakiri, M.; Guyeux, C.; Couchot, J.F.; Oudjida, A.K. Survey on hardware implementation of random number generators on FPGA: Theory and experimental analyses. *Comput. Sci. Rev.* **2018**, *27*, 135–153. [CrossRef]
18. Takada, A. White noise generation in a chaotic phase-locked loop. *IEICE Tech. Rep. Circuits Syst.* **2005**, *105*, 55–60.
19. Endo, T.; Chua, L. Chaos from Phase-Locked Loops. *IEEE Trans. Circuits Syst. I* **1988**, *35*, 987–1003. [CrossRef]
20. Endo, T.; Yokota, J. Generation of White Noise by Using Chaos in Practical Phase-Locked Loop Integrated Circuit Module. In Proceedings of the IEEE International Symposium on Circuits and Systems (ISCAS), New Orleans, LA, USA, 27–30 May 2007; pp. 201–204.

Article

Information Transfer Among the Components in Multi-Dimensional Complex Dynamical Systems

Yimin Yin and Xiaojun Duan *

College of Liberal Arts and Sciences, National University of Defense Technology, Changsha 410072, China;
yinyimin16@nudt.edu.cn
* Correspondence: xjduan@nudt.edu.cn; Tel.: +86-0731- 8700-1851

Received: 13 September 2018; Accepted: 7 October 2018; Published: 9 October 2018

Abstract: In this paper, a rigorous formalism of information transfer within a multi-dimensional deterministic dynamic system is established for both continuous flows and discrete mappings. The underlying mechanism is derived from entropy change and transfer during the evolutions of multiple components. While this work is mainly focused on three-dimensional systems, the analysis of information transfer among state variables can be generalized to high-dimensional systems. Explicit formulas are given and verified in the classical Lorenz and Chua's systems. The uncertainty of information transfer is quantified for all variables, with which a dynamic sensitivity analysis could be performed statistically as an additional benefit. The generalized formalisms can be applied to study dynamical behaviors as well as asymptotic dynamics of the system. The simulation results can help to reveal some underlying information for understanding the system better, which can be used for prediction and control in many diverse fields.

Keywords: Information transfer; continuous flow; discrete mapping; Lorenz system; Chua's system

1. Introduction

Uncertainty quantification in complex dynamical systems is an important topic in prediction models. By integrating information-theoretic methods to investigate potential physics and measure indices, the uncertainty can be quantified better in ensemble practical predictions of complex dynamical systems. For instance, one of the important motivations is the couplings among variables of dynamical systems generating information at a nonzero rate [1], which produces information exchange [2]. Entropy can be used to quantitatively describe production, gathering, exchange and transfer of information [3]. Information transfer analysis can be used to detect asymmetry in the interactions of subsystems [1,4]. The emergent phenomena cannot be simply derived or solely predicted from the knowledge of the structure or from the interactions among individual elements in complex systems [5]. The dynamics of information transportation plays a critical role in complex systems, resulting in the system prediction [6,7], controls of a system [8,9] and causal analysis [10,11]. It emphasizes further understanding and investigating information transportation in complex dynamical systems. It has been applied to quantify nonlinear interactions based on the information transfer by several underlying efficient estimation strategies in complex dynamical systems [12–14]. Simple examples are used to illustrate various complex phenomena. The formalisms about information transfer are mostly based on two time series [1,15–17].

Recently, a new approach on information flow between the components of two-dimensional (2D) systems was adapted by Liang and Kleeman [6], which can be used to deal with the change of the uncertainty of one component given by the other component. This idea is based on specific interactions between two components in complex dynamical systems. For a system with dynamics given, a measure of information transfer can be rigorously formulated (referred as LK2005 formalism henceforth in [6]).

In the forms of continuous flows and discrete mappings, the information flow has been analyzed using the Liouville equations [18] and the Frobenius–Perron operators [18]. These two equations are the evolution equations of the joint probability distributions, respectively. The present formalism is consistent with the transfer entropy of Schreiber [1] in both transfer asymmetry and quantification. A variety of generalizations and applications of the work in Reference [4] are developed in [19–25]. Majda and Harlim [26] applied the strategy to study subspaces of complex dynamical systems. For 2D systems, Liang and Kleeman discovered a concise law on the entropy evolution of deterministic autonomous systems and obtained the time rate of information flow from one component to the other [6]. Until now, the 2D formalism has been extended to some dynamical systems in different forms and scales with successful applications between two variables [23,25]. In the light of these applications, by thoroughly describing the statistical behavior of a system, this rigorous LK2005 formalism has yielded remarkable results [3].

However, the uncertainty of many real-world systems needs to be quantified among the variables for revealing the nonlinear relationships, so as to better understand the intrinsic mechanism and predict the forthcoming states of the systems [27]. Besides, many physical systems are affected by the interactions between multiple components in diverse fields [28]. For example, sensitivity analysis of an aircraft system with respect to design variables, parameters and uncertainty factors can be used to estimate the effects on the objective function or constraint function. The uncertainty analysis and sensitivity analysis (UASA) process is one of the key steps for determining the optimal search direction and guiding the design and decision-making, which aims at predicting complex computer models by quantifying the sensitivity information of the coupling variables. It can be offered to quick guide of determining design parameters which lead to high performance aircraft designs. Some preceding tools [29,30] related to sensitivity analysis are applicable for low-dimensional static problems and an urgent problem of high dimensionality arises when outputting variables of numerical models with spatially and temporally need to be solved [31]. The rigorous formalism of information flow has the potential to revolutionize the ability to analyze and measure uncertainty and sensitivity information in dynamical systems.

Hence, considering realistic applications, we generalize the LK2005 formalism to several variables of multi-dimensional dynamical systems in this paper. More precisely, we extend the results in [6,25] to the information flow between groups of components, rather than individual components. We aim to demonstrate that the formalism is feasible among several variables in arbitrary multi-dimensional dynamical systems when dynamics is fully known. In addition, the generalized formalisms can be reduced to two-dimensional formalisms as a special case. We also highlight the relationship between the LK2005 formalism and our generalized formalisms. Two applications are proposed with the classical Lorenz system and Chua's system as validations of our formalisms. Compared with the LK2005 formalism and the transfer mutual information method [32], the generalized formalisms are beneficial for revealing more information among variables. It can better explore the complexity of evolution and intrinsic regularity of multi-dimensional dynamical systems. Meanwhile, it can provide a simple and versatile method to analyze sensitivity in dynamical models. These generalized formulas enable one to understand the relationship between information transfer and the behavior of a system. It can be used to perform sensitivity analysis as a measure in multi-dimensional complex dynamical systems. Therefore, the generalized formalisms have much wider applications and are significant to investigate real-world problems.

The structure of this paper is as follows: Section 2 recalls a systematic introduction of the theories and the formalisms about information flow in 2D systems; In Section 3, the formalisms are generalized to adapt to multi-dimensional complex dynamical system components based on the LK2005 formalism. Details on the derivations of the formalisms and the related properties are demonstrated; Section 4 gives a description about the formalisms with multi-dimensional applications; the summary of this paper is given in Section 5.

2. Two-Dimensional Formalism of Information Transfer (the LK2005 Formalism [6])

2.1. Continuous Flows

For 2D continuous and deterministic autonomous systems with fully known dynamics,

$$\frac{d\mathbf{x}}{dt} = \mathbf{F}(\mathbf{x}), \tag{1}$$

where $\mathbf{F} = (F_1, F_2)$ with $F_i = F_i(x_1, x_2)$ for any $i = 1, 2$ is known as the flow vector and $\mathbf{x} = (x_1, x_2) \in \Omega = \Omega_1 \times \Omega_2$. A stochastic process $\mathbf{X} = (X_1, X_2) \in \Omega$ with joint probability density $\rho(x_1, x_2, t)$ at time t is the random variables corresponding to the sample values (x_1, x_2). For convenience, we will use the notation ρ or $\rho(x_1, x_2)$ instead of the notation $\rho(x_1, x_2, t)$ throughout Section 2, including the same expression at multi-dimensional cases in Section 3. In addition, the integral domain is the whole sample space Ω, except where noted. The probability density ρ associated with Equation (1) satisfies the Liouville equation [18]:

$$\frac{\partial \rho}{\partial t} + \frac{\partial(F_1 \rho)}{\partial x_1} + \frac{\partial(F_2 \rho)}{\partial x_2} = 0. \tag{2}$$

The rate of change of joint entropy of X_1 and X_2, $H(t) \overset{def}{=} - \iint_\Omega \rho \log \rho \, dx_1 dx_2$, satisfies the relation [6]

$$\frac{dH}{dt} = E(\nabla \cdot \mathbf{F}), \tag{3}$$

where E means the mathematical expectation with respect to ρ and $E(\nabla \cdot \mathbf{F}) = \iint_\Omega \rho(\nabla \cdot \mathbf{F}) dx_1 dx_2$. That is to say, when a system evolves with time, the change of its joint entropy is totally controlled by the contraction or expansion of the phase space [6]. Later on, Liang and Kleeman showed that this property holds for deterministic systems of arbitrary dimensionality [20].

Liang and Kleeman [6] provided a very efficient heuristic argument to describe the decomposition of the various evolutionary mechanisms of information transfer in terms of the individual and joint time rates of entropy changes of X_1, X_2 and (X_1, X_2). Firstly, they computed $\frac{dH_1}{dt}$ and $\frac{dH_2}{dt}$, where H_i is the entropy of X_i defined according to the marginal density, ρ_i. Secondly, they employed the novel idea of frozen variables to analyze the individual time rates of entropy changes. When X_i is fixed and X_j evolves on its own in 2D systems, they found its temporal rate of change of entropy depends only on $E(\frac{\partial F_j}{\partial x_j})$, denoted by $\frac{dH_j^*}{dt}$. In the presence of interactions between X_i and X_j, they observed that $\frac{dH_j}{dt} \neq E\left(\frac{\partial F_j}{\partial x_j}\right) = \frac{dH_j^*}{dt}$. Therefore, Liang and Kleeman [6] concluded that the difference between $\frac{dH_j}{dt}$ and $E\left(\frac{\partial F_j}{\partial x_j}\right)$ should equal to the rate of entropy transfer from X_i to X_j. In the meantime, they denoted the rate of flow from X_i to X_j by $T_{i \to j}$ (T stands for "transfer") and defined information flow/transfer as

$$T_{i \to j} = \frac{dH_j}{dt} - \frac{dH_j^*}{dt} = - \iint_\Omega \rho_{i|j}(x_i|x_j) \frac{\partial(F_j \rho_j)}{\partial x_j} dx_i dx_j, \tag{4}$$

where $\rho_{i|j}(x_i|x_j) = \frac{\rho(x_i, x_j, t)}{\rho(x_j, t)}$ and $i, j = 1, 2$ with different i, j at the same time.

2.2. Discrete Mappings

Similarly, Liang and Kleeman [6] also gave the formalism about a system in the discrete mapping form. Considering a 2D transformation

$$\Phi : \Omega \to \Omega, (x_1, x_2) \to (\Phi_1(\mathbf{x}), \Phi_2(\mathbf{x})),$$

where $\mathbf{x} = (x_1, x_2) \in \Omega$ and $\Omega := \Omega_1 \times \Omega_2$. The evolution of the density of Φ is driven by the Frobenius–Perron operator ($F - P$ operator) $P : L^1(\Omega) \to L^1(\Omega)$ [18]. The entropy increases as

$$\Delta H = -\iint P\rho \log P\rho \, dx_1 dx_2 + \iint \rho \log \rho \, dx_1 dx_2$$
$$= -\iint \rho(x_1, x_2) \log |J^{-1}| dx_1 dx_2,$$

where J^{-1} is the Jacobian matrix of Φ. When Φ_j is invertible in 2D transformations,

$$\Delta H_j^* = E \log |J_j|. \tag{5}$$

The entropy of X_j increases as

$$\Delta H_j = -\int_{\Omega_j} \left(\int_{\Omega_i} P\rho \, dx_i \right) \log \left(\int_{\Omega_i} P\rho \, dx_i \right) dx_j + \int_{\Omega_j} \rho_j \log \rho_j dx_j,$$

where ρ_j is the marginal density of X_j. When Φ_j is noninvertible in 2D transformations,

$$\Delta H_j^* = \int \rho_j(x_j) \log \rho_j(x_j) dx_j$$
$$- \iint P_j \rho_j \left(\Phi_j(x_i, x_j) \right) \log P_j \rho_j \left(\Phi_j(x_i, x_j) \right) \rho(x_i | x_j) |J_j| dx_i dx_j, \tag{6}$$

where P_j is the $F - P$ operator when x_i is frozen as a parameter in P_j. The entropy transferring from X_i to X_j is

$$T_{i \to j} = -\int_{\Omega_j} \left(\int_{\Omega_i} P\rho \, dx_i \right) \log \left(\int_{\Omega_i} P\rho \, dx_i \right) dx_j$$
$$+ \iint P_j \rho_j \left(\Phi_j(x_i, x_j) \right) \log P_j \rho_j \left(\Phi_j(x_i, x_j) \right) \rho(x_i | x_j) |J_j| dx_i dx_j, \tag{7}$$

where $i, j = 1, 2$ with different i, j at the same time.

3. n-Dimensional Formalism of Information Transfer

3.1. Continuous Flows

Firstly, we consider a three-dimensional (3D) continuous autonomous system,

$$\frac{d\mathbf{x}}{dt} = \mathbf{F}(\mathbf{x}), \tag{8}$$

where $\mathbf{F} = (F_1, F_2, F_3)$ is a known flow vector. Similarly, the probability density ρ associated with Equation (8) satisfies the Liouville equation [18]:

$$\frac{\partial \rho}{\partial t} + \frac{\partial (F_1 \rho)}{\partial x_1} + \frac{\partial (F_2 \rho)}{\partial x_2} + \frac{\partial (F_3 \rho)}{\partial x_3} = 0. \tag{9}$$

Analogous to the derivation in [6], firstly, multiplying by $(1 + \log \rho)$ for Equation (9), after some algebraic manipulations:

$$\frac{\partial (\rho \log \rho)}{\partial t} + \mathbf{F} \cdot (\nabla \cdot (\rho \log \rho)) + \rho(1 + \log \rho)\nabla \cdot \mathbf{F} = 0. \tag{10}$$

Then, integrating for Equation (10),

$$\frac{dH}{dt} - \iiint_{\Omega} \nabla \cdot (\rho \log \rho \mathbf{F}) dx_1 dx_2 dx_3 - \iiint_{\Omega} \rho \nabla \cdot \mathbf{F} dx_1 dx_2 dx_3 = 0.$$

Assuming that ρ vanishes at the boundaries (the compact support assumption for ρ and the assumption is reasonable in real-world problems [6]), it is found that the time rate of the joint entropy change of X_1, X_2 and X_3,

$$H(t) \overset{def}{=} - \iiint_{\Omega} \rho \log \rho dx_1 dx_2 dx_3,$$

satisfies

$$\frac{dH}{dt} - \iiint_{\Omega} \rho(x_1, x_2, x_3) \nabla \cdot \mathbf{F} dx_1 dx_2 dx_3 = 0$$

or

$$\frac{dH}{dt} = E(\nabla \cdot \mathbf{F}),$$

where $E(\nabla \cdot \mathbf{F}) = \iiint_{\Omega} \rho(\nabla \cdot \mathbf{F}) dx_1 dx_2 dx_3$.

As mentioned above, the time rate of change of H equals to the mathematical expectation of the divergence of the flow vector $\mathbf{F}$. When we are interested in the entropy evolution of a component, x_k in 3D systems, the marginal density is

$$\rho_k(x_k, t) = \iint_{\Omega_i \times \Omega_j} \rho(x_i, x_j, x_k, t) dx_i dx_j.$$

The evolution equation of ρ_k is derived by taking the integral of Equation (9) with respect to x_i and x_j over the subspace $\Omega_i \times \Omega_j$:

$$\frac{\partial \rho_k}{\partial t} + \frac{\partial}{\partial x_k} \iint_{\Omega_i \times \Omega_j} \rho F_k dx_i dx_j = 0.$$

The third and fourth terms in Equation (9) have been integrated out with the compact support assumption for ρ. So the entropy for the component

$$H_k(t) = - \int_{\Omega_k} \rho_k \log \rho_k dx_k$$

evolves as

$$\frac{dH_k}{dt} = \iiint_{\Omega} \left[\log \rho_k \frac{\partial(\rho F_k)}{\partial x_k} \right] dx_i dx_j dx_k,$$

i.e.,

$$\frac{dH_k}{dt} = - \iiint_{\Omega} \rho \left[\frac{F_k}{\rho_k} \frac{\partial \rho_k}{\partial x_k} \right] dx_i dx_j dx_k. \tag{11}$$

The Equation (11) states how H_k evolves with time. The evolutionary mechanism of H_k derives from two parts: One is from the evolution itself, $\frac{dH_k^*}{dt}$; another from the transfers of X_i and X_j according to the coupling in the joint density distribution ρ. From Section 2, we know that when X_k evolves on its own, then

$$E\left(\frac{\partial F_k}{\partial x_k} \right) = \frac{dH_k^*}{dt} = \iiint_{\Omega} \rho \frac{\partial F_k}{\partial x_k} dx_i dx_j dx_k.$$

Therefore, the rate of information flow/transfer from X_i, X_j to X_k is

$$
\begin{aligned}
T_{i,j\to k} &= \frac{dH_k}{dt} - \frac{dH_k^*}{dt} = \iiint_\Omega \rho \left(\frac{F_k}{\rho_k}\frac{\partial \rho_k}{\partial x_k} + \frac{\partial F_k}{\partial x_k} \right) dx_i dx_j dx_k \\
&= -\iiint_\Omega \frac{\rho}{\rho_k} \frac{\partial (F_k \rho_k)}{\partial x_k} dx_i dx_j dx_k \\
&= -\iiint_\Omega \rho_{i,j|k}(x_i, x_j | x_k) \frac{\partial (F_k \rho_k)}{\partial x_k} dx_i dx_j dx_k,
\end{aligned}
\tag{12}
$$

where $\rho_{i,j|k}(x_i, x_j | x_k) = \frac{\rho(x_i, x_j, x_k, t)}{\rho(x_k, t)}$ and $i, j, k = 1, 2, 3$ with different i, j, k at the same time.

In particular, if $F_1 = F_1(x_1)$ has no dependence on x_2, then $T_{2\to 1} = 0$. There is no information transfer from random variable component X_2 to X_1. This holds true with the transfers defined in LK2005 formalism. Obviously, in system (8), when F_1 has no dependence on x_2, x_3, there should be no information transfer from X_2, X_3 to X_1, but there is possibility that the transfers in other directions may be nonzero when F_2 depends on x_1, x_3 or F_3 depends on x_1, x_2. This is consistent with the information transfer defined in Equation (12). As a matter of fact, an important property of the transfer is given below.

Theorem 1. *If F_k is independent of x_i, x_j in system (8) with different i, j, k, then $T_{i,j\to k} = 0$.*

Proof of Theorem 1. According to the formalism of information transfer for system (8), with the notation of $F_k = F_k(x_k)$,

$$
\begin{aligned}
T_{i,j\to k} &= -\iiint_\Omega \rho_{i,j|k}(x_i, x_j | x_k) \frac{\partial (F_k \rho_k)}{\partial x_k} dx_i dx_j dx_k \\
&= -\int_{\Omega_k} \left(\iint_{\Omega_i \times \Omega_j} \rho_{i,j|k}(x_i, x_j | x_k) dx_i dx_j \right) \frac{\partial (F_k \rho_k)}{\partial x_k} dx_k \\
&= -\int_{\Omega_k} \frac{\partial (F_k \rho_k)}{\partial x_k} dx_k = 0.
\end{aligned}
$$

$\square$

It is worth noting that, while X_k gains information from X_i or X_j, or X_i and X_j, X_i or X_j might have no dependence on X_k in 3D systems. An important property about information transfer is its asymmetry among the components [1]. In addition, it is interesting to note that the formalism of 3D systems can be reduced to 2D cases under the condition that one variable does not depend on another variable. For example, If the evolution of X_k is independent of X_i, then

$$
\begin{aligned}
T_{i,j\to k} &= -\iiint_\Omega \rho_{i,j|k}(x_i, x_j | x_k) \frac{\partial (F_k \rho_k)}{\partial x_k} dx_i dx_j dx_k \\
&= -\iint_{\Omega_j \times \Omega_k} \left(\int_{\Omega_i} \rho(x_i, x_j, x_k) dx_i \right) \frac{1}{\rho(x_k)} \frac{\partial (F_k \rho_k)}{\partial x_k} dx_j dx_k \\
&= -\iint_{\Omega_j \times \Omega_k} \rho(x_j | x_k) \frac{\partial (F_k \rho_k)}{\partial x_k} dx_j dx_k = T_{j\to k}.
\end{aligned}
\tag{13}
$$

In particular, when X_k is independent of X_i and X_j,

$$
T_{i,j\to k} = T_{i\to k} = T_{j\to k} = 0.
$$

According to Theorem 1, the results are apparent. Furthermore, when X_k depends on X_i and X_j,

$$
\begin{aligned}
T_{i,j \to k} &= -\iiint_\Omega \rho_{i,j|k}(x_i, x_j | x_k) \frac{\partial(F_k \rho_k)}{\partial x_k} dx_i dx_j dx_k \\
&= -\iiint_\Omega \rho(x_j | x_k) \cdot \rho_{i|j,k}(x_i | x_j, x_k) \frac{\partial(F_k \rho_k)}{\partial x_k} dx_i dx_j dx_k \\
&= -\int_{\Omega_i} \left(\iint_{\Omega_j \times \Omega_k} \rho_{j|k}(x_j | x_k) \frac{\partial(F_k \rho_k)}{\partial x_k} dx_j dx_k \right) \rho_{i|j,k}(x_i | x_j, x_k) dx_i \\
&= -\int_{\Omega_i} T_{j \to k} \cdot \rho_{i|j,k}(x_i | x_j, x_k) dx_i,
\end{aligned}
$$

or

$$
T_{i,j \to k} = -\int_{\Omega_j} T_{i \to k} \cdot \rho_{j|i,k}(x_j | x_i, x_k) dx_j. \tag{14}
$$

From the above derivations, we can see that our formalisms are further intensified by emphasizing the inherent relation with the formalisms in 2D systems. The information flows from two variables and the high order interactions between them to another variable are quantified by formula (12). These are generalized forms of the LK2005 formalism. In Section 4, we will validate the conclusions by the applications of all formulas in the Lorenz and Chua's systems. Moreover, when several variables are involved, the formalisms are capable to tackle information transfers of a multi-dimensional system.

Combining the Liouville equation

$$
\frac{\partial \rho}{\partial t} + \frac{\partial(F_1 \rho)}{\partial x_1} + \frac{\partial(F_2 \rho)}{\partial x_2} + \cdots + \frac{\partial(F_n \rho)}{\partial x_n} = 0, \tag{15}
$$

with Equation (3), $\frac{dH}{dt} = E(\nabla \cdot \mathbf{F})$ in n-dimensional situations, we can generalize the formalism to n-dimensional continuous and deterministic autonomous systems in the same way. For example, the transfer of information from components $X_2, X_3, \ldots, X_n$ to X_1 is

$$
T_{2,3,\ldots,n \to 1} = -\int_\Omega \rho_{2,3,\ldots,n}(x_2, x_3, \ldots, x_n | x_1) \frac{\partial(F_1 \rho_1)}{\partial x_1} dx_1 dx_2 \ldots dx_n.
$$

Hence, Theorem 1 can be generalized to multi-dimensional cases.

3.2. Discrete Mappings

For a 3D transformation $\Phi : \Omega \to \Omega, (x_1, x_2, x_3) \to (\Phi_1(\mathbf{x}), \Phi_2(\mathbf{x}), \Phi_3(\mathbf{x}))$, the evolution of its density is driven by the Frobenius–Perron operator ($F - P$ operator) $P : L^1(\Omega) \to L^1(\Omega)$ [18]. Similar to the 2D case, after some efficient computations, the entropy transfer from X_i, X_j to X_k in three-dimensional mappings has the following form:

$$
\begin{aligned}
T_{i,j \to k} =& \Delta H_k - \Delta H_k^* \\
=& -\int_{\Omega_k} \left(\iint_{\Omega_i \times \Omega_j} P\rho dx_i dx_j \right) \log \left(\iint_{\Omega_i \times \Omega_j} P\rho dx_i dx_j \right) dx_k \\
& + \iiint P_k \rho_k \left(\Phi_k(x_i, x_j, x_k) \right) \log P_k \rho_k \left(\Phi_k(x_i, x_j, x_k) \right) \rho(x_i, x_j | x_k) |J_k| dx_i dx_j dx_k.
\end{aligned} \tag{16}
$$

We also give a theorem for the discrete mappings and highlight the relationship between two-dimensional formalisms and generalized formalisms. The formalisms can be extended to high-dimensional situations as well. The detailed processes are demonstrated in Appendix A.

4. The Application of Multi-Dimensional Formalism of Information Transfer

4.1. The Lorenz System

In this section, we propose an application to study the information flows about the Lorenz system [33]:

$$
\begin{cases}
\dfrac{dx_1}{dt} = \sigma(x_2 - x_1) \\[4pt]
\dfrac{dx_2}{dt} = x_1(r - x_3) - x_2 \\[4pt]
\dfrac{dx_3}{dt} = x_1 x_2 - b x_3
\end{cases}
,
$$

where σ, r and b are parameters, x_1, x_2 and x_3 are the system state variables, and t is time. A chaotic attractor of Lorenz system with $\sigma = 10, r = 28, b = \frac{8}{3}$ is shown in Figure 1.

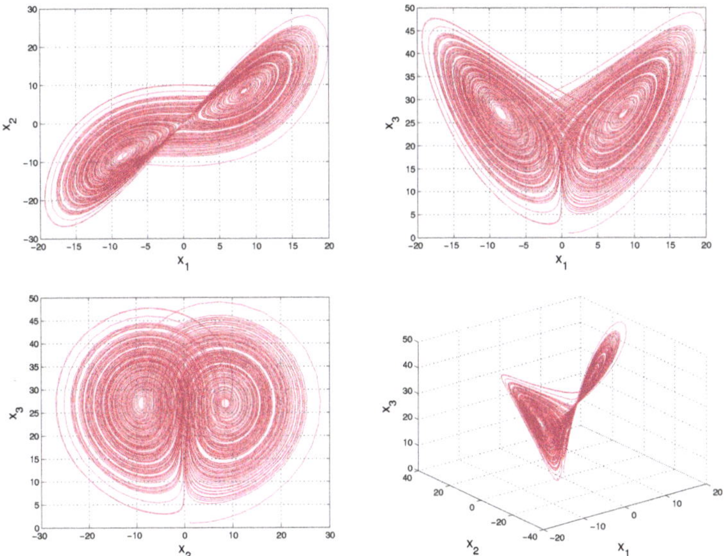

Figure 1. The Lorenz attractor with initial value (1,1,1).

Firstly, we need to obtain the joint probability density function $\rho(x_1, x_2, x_3)$ of **X** to calculate information flows among the variables. For a deterministic system with known dynamics, the underlying evolution of the joint density $\rho(x_1, x_2, x_3)$ can be obtained by solving the Liouville equation. Taking into account of the computational load, we estimate the joint density $\rho(x_1, x_2, x_3)$ via numerical simulations. The steps are summarized as follows:

- Initialize the joint density $\rho(x_1, x_2, x_3)$ with a preset distribution ρ_0, then generate an ensemble through drawing samples randomly according to the initial distribution ρ_0.
- Partition the sample space Ω into "bins".
- Obtain an ensemble prediction for the Lorenz system at every time step.
- Estimate the three-variable joint probability density function ρ via bin counting at every time step.

The Lorenz system is solved by applying a fourth order Runge–Kutta method with a time step $\Delta t = 0.01$. According to Figure 1, the computation domain is restricted to $\Omega \equiv [-30, 30] \times [-30, 30] \times [0, 60]$, which includes the attractor of the Lorenz system. We discretize the sample space into $60 \times 60 \times 60 = 21,600$ bins to ensure covering the whole attractor and one draw per bin on average via making 21,600 random draws. Initially, we assume **X** is distributed as a Gaussian process $N(u(t), \Sigma(t))$, with a mean u and a covariance matrix Σ:

$$u(0) = \begin{bmatrix} u_1 \\ u_2 \\ u_3 \end{bmatrix}, \Sigma(0) = \begin{bmatrix} \sigma_1^2 & 0 & 0 \\ 0 & \sigma_2^2 & 0 \\ 0 & 0 & \sigma_3^2 \end{bmatrix}.$$

Although we have used different parameters u and σ_d^2 ($d = 1, 2, 3$) to compute information flows for the Lorenz system, the final results are the same and the trends stay invariant. The parameters u and σ_d^2 can be adjusted for different experiments. Here we only show the results of one experiment with $u_d = 4$ and $\sigma_d^2 = 4$. The ensemble is developed by drawing sample randomly in the light of a pre-established distribution $\rho_0(\underline{x})$. We obtain an ensemble of **X** and estimate the three-variable joint probability density function $\rho(x_1, x_2, x_3, t)$ by the way of counting the bins, at every time step. As the equations are integrated forward in the Lorenz system, ρ can be estimated as a function of time and describe the statistics of the system. A detailed discussion on probability estimation through bin counting are referred to [20,25]. The sample data with initial value (1,1,1) and an estimated marginal density of x_1, x_2 and x_3 are displayed in Figure 2.

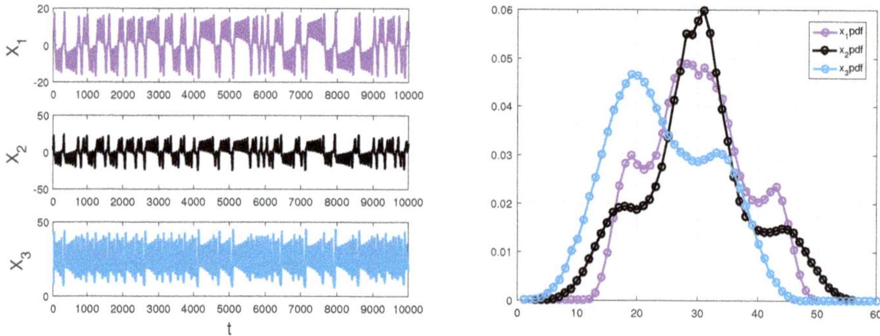

Figure 2. Left panel: a sample data (X_1, X_2 and X_3) of the Lorenz system generated by a fourth order Runge–Kutta method with $\Delta t = 0.01$. Right panel: an estimated marginal density of x_1, x_2 and x_3 via counting the bins and initializing with a Gaussian distribution, respectively.

Through formula (12), the information transfer within three variables can be computed. There are nine transfer series in the Lorenz system, but here we mainly focus on the couple effect from two components to another component, that is, $T_{i,j \to k}, i, j, k = 1, 2, 3$ with different i, j, k at the same time. A nonzero $T_{i,j \to k}$ means that X_i and X_j are causal to X_k, and the value means how much uncertainty that X_i and X_j bring to X_k. Among all the transfers, it is clearly shown that any two variables drive the other variable in the dynamics except the evolution of X_1 which only depends on X_2. For the sake of revealing some underlying information in the chaotic dynamical system better, we also give information transfer among the components over space with a Gaussian distribution initialization and the averaged density over time via using the following formula: $S_{i,j \to k} = -\iint \bar{\rho}_{i,j|k}(x_i, x_j|x_k, t) \frac{\partial (F_k \bar{\rho}_k(t))}{\partial x_k} dx_i dx_j$, which characterizes the strength of information transfer at different planes of $x = x_i$. That is to say, it demonstrates the information transfer of x_j and x_k to x_i plane, whose relative values represent the magnitudes of information transfer. The calculation results are plotted in the left panel and the right one of Figure 3, respectively. According to the magnitude of parameters in the Lorenz system and the definition of rigorous 3D formalisms, the information transfer from X_1, X_2 to X_3 is the smallest. The results are just as we expected, $|T_{1,2 \to 3}| < |T_{1,3 \to 2}| < |T_{2,3 \to 1}|$, as shown in the left panel of Figure 3. Meanwhile, we can get much information through numerical simulations. For example, the information transfer from X_2 and X_3 on X_1 is larger than that of X_1 and X_3 on X_2 in the Lorenz system, which is helpful for us to better analyze the system and the fields of interest. Only the absolute value of T measures the information transfer among the variables [23]. As the ensemble evolution is carried forth, any two variables aim to reduce the uncertainty of the other variable [24]; in other words, any two variables tend to stabilize the other variable. All information flows go to constants, which means that the system tends to be stable simultaneously. Comparing the left panel with the right one in Figure 3, we can find that not only the information flow from X_2 and X_3 to X_1 is the largest at different times, but also the total information transfer is the largest at x_1 plane, and the strength of information transfer obeys a distribution in each direction of x. Repeated experiments are found to be in line with the results no matter whatever the initialization is given.

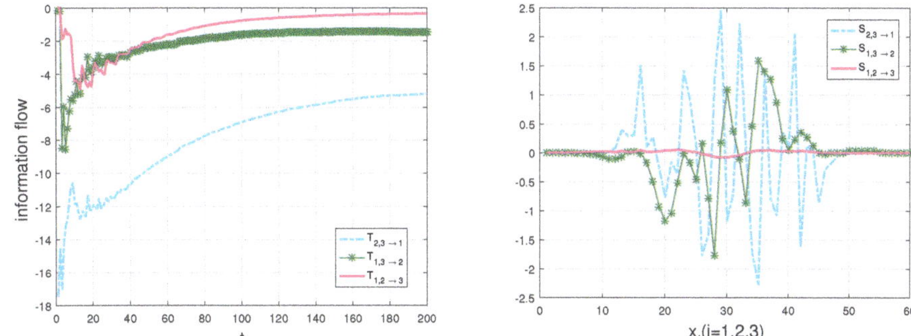

Figure 3. Left panel: the multivariate information flow of the Lorenz system: blue dot-dash line: $T_{2,3\rightarrow1}$; green star line: $T_{1,3\rightarrow2}$; red solid line: $T_{1,2\rightarrow3}$ (in nats per unit time); Right panel: the information strength of transfer in the Lorenz system: blue dot-dash line: $S_{2,3\rightarrow1}$; green star line: $S_{1,3\rightarrow2}$; red solid line: $S_{1,2\rightarrow3}$ (arbitrary unit).

In particular, we compute the transfer, $T_{2\rightarrow1}$, then compare $T_{2\rightarrow1}$ with the transfer, $T_{2,3\rightarrow1}$ in Figure 4, as well as plot the transfers $T_{1\rightarrow2}$, $T_{3\rightarrow2}$ and $T_{1,3\rightarrow2}$ in Figure 5.

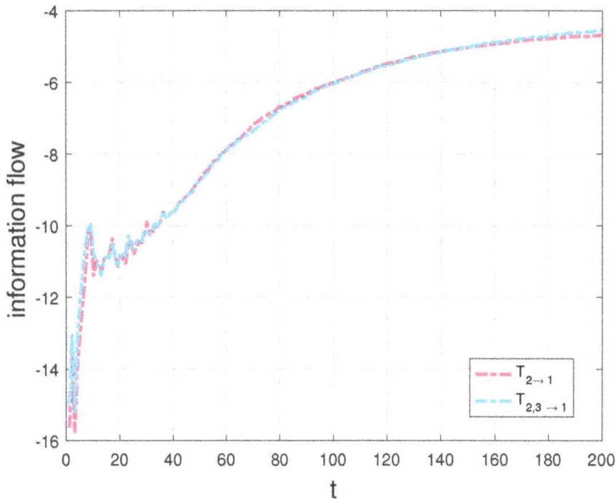

Figure 4. $T_{2\rightarrow1}$ and $T_{2,3\rightarrow1}$ in the Lorenz system (in nats per unit time).

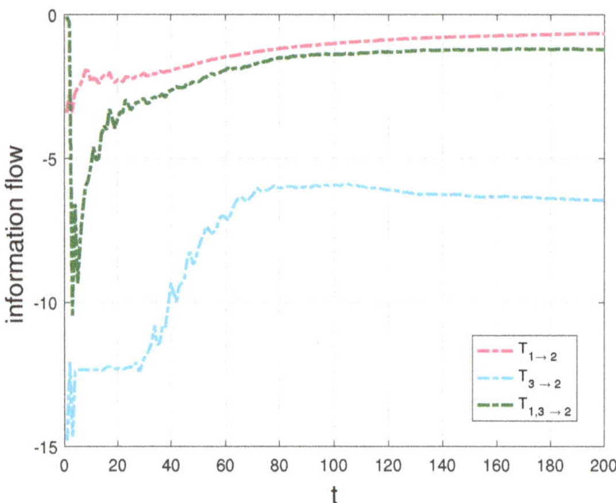

Figure 5. $T_{1\to2}$, $T_{3\to2}$ and $T_{1,3\to2}$ in the Lorenz system (in nats per unit time).

Since the evolution of X_1 is independent of X_3 and the evolution of X_2 depends on X_1 and X_3 in the Lorenz system, the transfer $T_{2\to1}$ should be equal to $T_{2,3\to1}$ and neither the transfer $T_{1\to2}$ nor $T_{3\to2}$ should not be equal to $T_{1,3\to2}$ according to the derivations in Section 3.1. As expected, there is almost no difference between the two flows in Figure 4. The interpretation of the results is that X_3 is not causal to X_1 in the Lorenz system. The result agrees well with theoretical analysis, which also validates our formalisms. But the graphs $T_{1\to2}$ and $T_{3\to2}$ are quite different from the graph $T_{1,3\to2}$ in Figure 5, as that both X_1 and X_3 are causal to X_2 in the Lorenz system. From Figures 4 and 5, we can find that the information flow $T_{2\to1}$ is different from $T_{1\to2}$, as a property of asymmetry of the information transfer. There exists hidden sensitivity information in information transfer processes of high-dimensional dynamical systems: whether or not one variable brings more uncertainty to another variable. Comparing the magnitudes of three flows in Figure 5, we can say that X_3 is more sensitive to X_2 than X_1 to X_2 from the sensitivity analysis point of view. All the above differences are exactly the embodiment of the differences between the information flows in multi-dimensional dynamical systems and the LK2005 formalism. The proposed formalisms can be used to measure information transfers among the variables in dynamical systems and the numerical results can show how the measurement behavior with time, compared with the qualification of information transfer between two variables [4] and the transfer mutual information method [32]. For example, it can be quantified the influence that x_3 on the relationship between x_1 and x_2 using the transfer mutual information method in the Lorenz system. With our generalized formalisms, we can give quantitatively the influence from x_3 to the relationship between x_1 and x_2 as a dynamical process and other relationships (such as the asymmetrical influence between two variables) among the variables for analyzing the system better. To test the influence of error propagation on the measurement of information transfers, we use a different natural interval extension to compute information transfers according to the striking method [34]. In other words, we compute information transfers using formula (12) in the Lorenz system with the rewritten second equation, that is, $rx_1 - x_1x_3$ is used to replace $x_1(r - x_3)$. For the Lorenz system, the results show that, the algorithm performs well (the relative error $< 2\%$). All simulations are performed on a 64-bit Matlab R2016a environment. The physical consistency of the proposed approach in this paper can be explained as that a direction of the phase space is frozen in order to extract information transfers from the other two directions [3]. In addition, nonlinearity may lead a deterministic system to chaos, which causes the "spikes" in the right panel of Figure 3 and corresponds to intermittent switching in the chaotic dynamics. As the remarkable theory stated in [35], it indicates when the dynamics are about to switch lobes of the attractor in the Lorenz system.

Since Liouville equations and Frobenius–Perron analysis describe an ensemble of trajectories, we can use the generated formalisms of information flow as a sensitivity analysis index to perform dynamic sensitivity information analysis instead of the preceding widely used methods such as repeated calculation of principal component coefficients [36,37], construction of functional metamodels [31,38], calculation of moving average of the sensitivity index [39] and direct perturbation analysis of a dynamical system [40]. Using information flow to identify sensitive variables is directly based on the statistical perspective, which can improve numerical accuracy and efficiency while reduce the calculation load, compared with conventional dynamic sensitivity analysis methods. We cannot only quantify how much the uncertainty among variables of a system, but also understand how they influence system behavior, so it may be measured and used for prediction and control in realistic applications.

Furthermore, we use Equation (15) to compute information transfers, $T_{yzw \to x}$, $T_{xzw \to y}$, $T_{xyw \to z}$, and $T_{xyz \to w}$ with the same strategy in the four-dimensional (4D) dynamical system:

$$\begin{cases} \dfrac{dx}{dt} = 12(y - x) \\ \dfrac{dy}{dt} = 23x - xz - y + w \\ \dfrac{dz}{dt} = xy - 2.1z \\ \dfrac{dw}{dt} = -6y - 0.2w \end{cases},$$

whose results are shown in Figure 6.

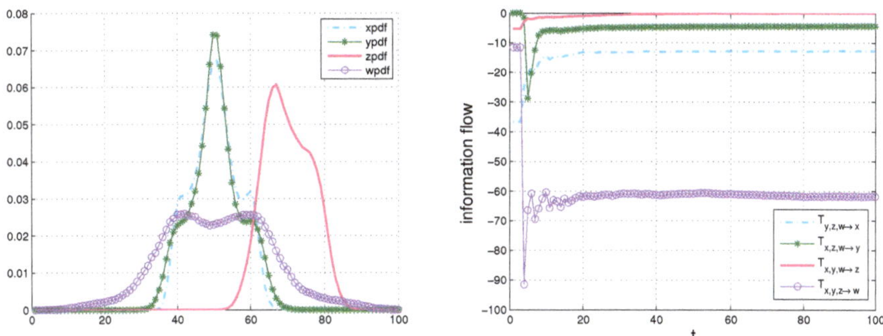

Figure 6. Left panel: an estimated marginal density of x, y, z and w via counting the bins and initializing with a Gaussian distribution, respectively; Right panel: the multivariate information flow over time of a 4D dynamical system.

The generalized formalisms are useful to deal with universal problems, which is not difficult to be applied to higher-dimensional cases.

4.2. The Chua's System

Since it is the first analog circuit to realize chaos in experiments, the initial Chua's system is a well-known dynamical model [41]. The Chua's system is described in reference [42] and there are many researches on its dynamical behavior [43,44]. Here we present an investigation of the information flows within the smooth Chua's system [45]:

$$\begin{cases} \dfrac{dx}{dt} = p(x + y - x \ln \sqrt{1 + x^2}) \\ \dfrac{dy}{dt} = x - y + z \\ \dfrac{dz}{dt} = -qy \end{cases},$$

where p, q are parameters, x, y and z are state variables in $\mathbf{R}$ and $t \in \mathbf{R}^+$. When $p = 11$ and $q = 14.87$, a chaotic attractor of the Chua's system is shown in Figure 7.

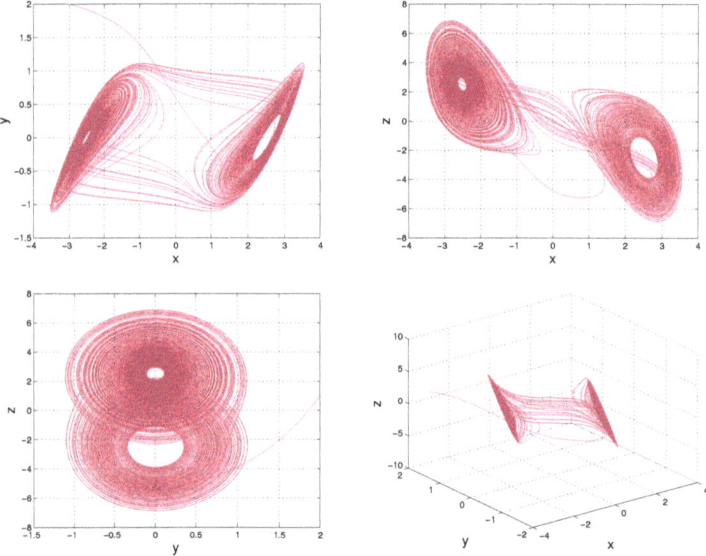

Figure 7. The attractor of Chua's system with $x(0) = -3, y(0) = 2, z(0) = 1$. The former three trajectories are x, z-plane, x, y-plane and y, z-plane, respectively. The last trajectory is a 3D plot of x, y and z.

As mentioned before, using the same estimation procedures, we can obtain the density $\rho(x, y, z)$ of $\mathbf{R}$ by counting the bins at each step. From Figure 7, the appropriate computation domain $\Omega \equiv [-10, 10] \times [-10, 10] \times [-10, 10]$ which includes an attractor of the Chua's system can be selected to estimate the three-variable joint probability density function. The following computation is demonstrated by applying a fourth order Runge–Kutta method. Similarly, we only show the results of one experiment after computing information flows multiple times by using different parameters. Suppose that $\mathbf{R}$ is distributed as a Gaussian process $N(u(t), \Sigma(t))$, with a mean u and a covariance matrix Σ in the initial state:

$$u(0) = \begin{bmatrix} 9 \\ 9 \\ 9 \end{bmatrix}, \Sigma(0) = \begin{bmatrix} 9 & 0 & 0 \\ 0 & 9 & 0 \\ 0 & 0 & 9 \end{bmatrix}.$$

Due to the additional fact that the smooth Chua's circuit has a highly non-coherent dynamics [46], we discretize the sample space into $200 \times 200 \times 200 = 8,000,000$ bins to adequately understand the information transfer and the behavior of the system over time. A sample data and an estimation result of three marginal densities are shown in Figure 8, and we can find that the dynamical behaviors of the system are consistent with the results, such as symmetry. Using formula (12) to compute the information transfers within three variables of Chua's system. Firstly, we discuss the coupling effect from two components to the other component, the calculation results are demonstrated in Figure 9.

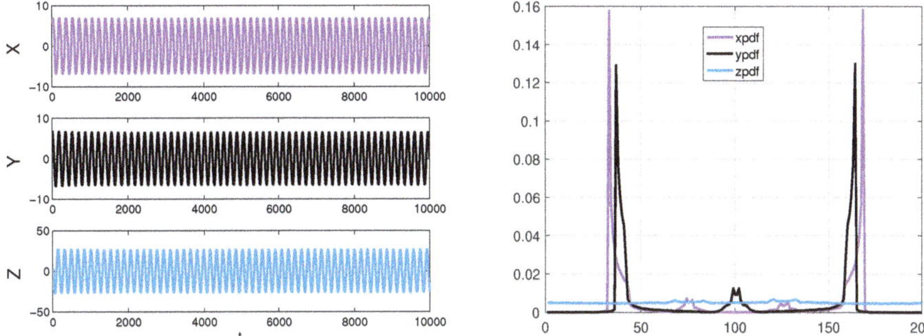

Figure 8. Left panel: a sample data (X, Y and Z) of the Chua's system generated by a fourth order Runge–Kutta method with $\Delta t = 0.01$; Right panel: the purple line, black line, and blue line represent an estimated marginal density of x, y, z by counting bins, respectively.

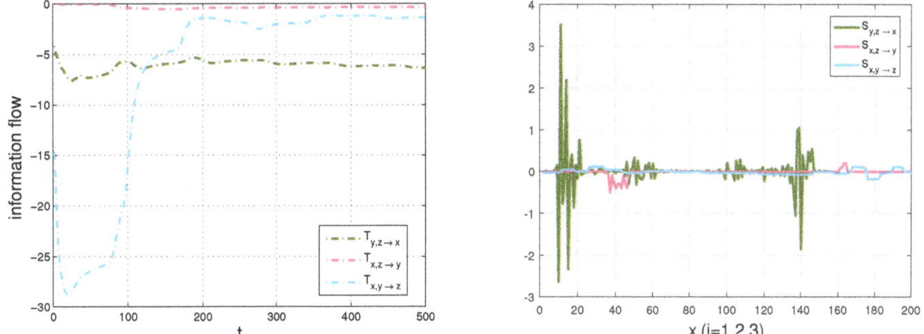

Figure 9. Left panel: the multivariate information flow of the Chua's system: green dot-dash line: $T_{y,z \to x}$; red dot-dash line: $T_{x,z \to y}$; blue dot-dash line: $T_{x,y \to z}$ (in nats per unit time); Right panel: the information strength of transfer in the Chua's system: green dot-dash line: $S_{y,z \to x}$; red dot-dash line: $S_{x,z \to y}$; blue dot-dash line: $S_{x,y \to z}$ (arbitrary unit).

Secondly, we compute the transfers, $T_{y \to z}$ and $T_{z \to y}$, then compare $T_{y \to z}$ with the transfer, $T_{x,y \to z}$ and $T_{z \to y}$ with $T_{x,z \to y}$ in Figures 10 and 11, respectively. We also show the corresponding results of the strength of information transfer among the components with a Gauss distribution initialization and the averaged density over time in Figure 9.

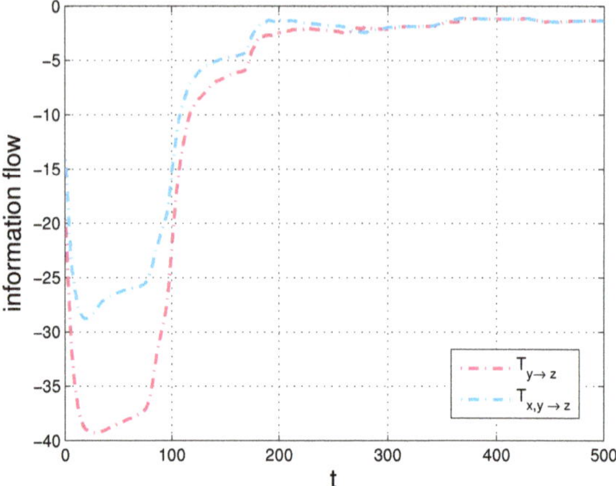

Figure 10. $T_{y \to z}$ and $T_{x,y \to z}$ in the Chua's system (in nats per unit time).

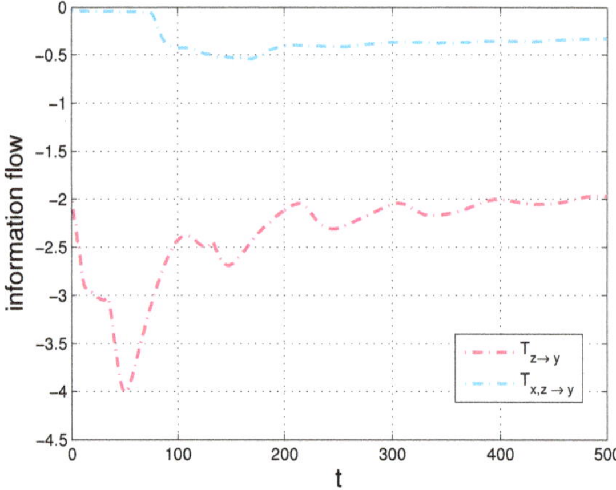

Figure 11. $T_{z \to y}$ and $T_{x,z \to y}$ in the Chua's system (in nats per unit time).

Since X causes Y but does not cause Z in the Chua's system, the numerical results of Figures 10 and 11 conform with the derivations of Equations (13) and (14) in Section 3.1. More specifically, there is almost no difference between the two flows in Figure 10, however, there exists large disparity between the two flows in Figure 11. The results also verify our formalisms. In addition, as shown in Figures 10 and 11, we can see that the information flow $T_{y \to z}$ is different from $T_{z \to y}$ due to the asymmetry of information transfer. All simulations are performed on a 64-bit Matlab R2016a environment. We are able to estimate that one variable makes another variable more uncertain or more predictable via the generalized formalisms. Besides, we can identify sensitive variables by computing information transfers among the variables in dynamical systems.

Compared with the Lorenz system, the Chua's system embodies in engineering systems besides that their discoveries were extraordinary and changed scientific thinking [46]. It can be used as another means to research, experiment and think about humanity, identity and art, etc. [47,48]. In studying visualization of the dynamics of Chua's circuit through computational models, the quantitative transformations of behavior are being taken into account [46]. The multi-dimensional formalisms of information flow enable us to improve our ability to estimate, predict, and control complex systems in many diverse fields. Furthermore, most existing approaches in control and synchronization of chaotic systems require adjusting the parameters of the model and estimating system parameters, which become an active area of research [49], and an additional benefit provided by the multi-dimensional formalisms of information flow is parameter estimation. We can compute information flows of the simulation model with different sets of parameters and do the same procedure for obtaining a group of feedback, then determine the optimal parameters that cater for the actual needs in order to put insight into complex behavior of models by comparing the change rates.

5. Conclusions

Based on the LK2005 formalism, we propose a rigorous and general formalism of the information transfer among multi-dimensional complex dynamical system components, for continuous flows and discrete mappings, respectively. Information transfers are quantified through entropy transfers from some components to another component, enabling us to better understand the physical mechanism underlying the superficial behavior and explore deeply hidden information in the evolution of multi-dimensional dynamical systems. When the generalized formalisms are reduced to 2D cases, the results are consistent with the LK2005 formalism. We mainly focus on the study of 3D systems and apply the formalisms to investigate information transfers for the Lorenz system and the Chua's system. In the above-mentioned two cases, we show that information flows of the whole evolution and the strength of information transfer at different planes, which implies that how uncertainty propagates and how dynamic essential information in the system transports. The results of experiments on the generalized formalisms conform with observations and empirical analysis in the literature, whose application may benefit many diverse fields. Compared with the qualification of information transfer between two variables [4] and the transfer mutual information method [32], the generalized formalisms are helpful for analyzing the relationships among the variables in dynamical systems and the research of complex systems. Moreover, since the formalism is built on the statistical nature of information, it has the potential to perform sensitivity analysis in multi-dimensional complex dynamical systems and advance our ability to estimate, predict and control these systems. In practice, for complex high-dimensional dynamical systems, it is not easy to give the dynamics analytically. Considering many critical data-driven problems are primed to take advantage of progress in the data-driven discovery of dynamics [35], we are developing a dynamic-free formulation to analyze information flows of multi-dimensional dynamical systems.

In the future, the formalism will be further generated to high-dimensional stochastic dynamical systems and time-delay systems. Meanwhile, future research should investigate how the information flow as a new indicator can be deployed in the frame of dynamic sensitivity analysis.

Author Contributions: Y.Y. proposed the original idea, implemented the experiments, analyzed the data and wrote the paper. X.D. contributed to the theoretical analysis and simulation designs and revised the manuscript. All authors read and approved the manuscript.

Acknowledgments: This work is supported by the National Natural Science Foundation of China (Grant No. 11771450).

Appendix A. Discrete Mappings

Now consider a 3D transformation

$$\Phi : \Omega \to \Omega, (x_1, x_2, x_3) \to (\Phi_1(\mathbf{x}), \Phi_2(\mathbf{x}), \Phi_3(\mathbf{x})) \tag{A1}$$

and the Frobenius–Perron operator ($F - P$ operator) $P : L^1(\Omega) \to L^1(\Omega)$ [18] which steers the evolution of its density. Loosely, given a density $\rho = (x_1, x_2, x_3)$, P is defined as that

$$\iiint_w P\rho(x_1, x_2, x_3) dx_1 dx_2 dx_3 = \iiint_{\Phi^{-1}(w)} \rho(x_1, x_2, x_3) dx_1 dx_2 dx_3,$$

where w represents any subset of Ω. When Φ is invertible, P can be expressed clearly as $P\rho(\underline{x}) = \rho\left[\Phi^{-1}(\underline{x})\right] |J^{-1}|$, where $J^{-1} = J^{-1}(x_1, x_2, x_3) = \det\left[\frac{\partial(\Phi^{-1}(x_1,x_2,x_3))}{\partial(x_1,x_2,x_3)}\right]$ is the determinant of the Jacobian matrix of Φ. Similar to the two-dimensional case, the entropy increases

$$\begin{aligned}
\Delta H &= -\iiint P\rho \log P\rho \, dx_1 dx_2 dx_3 + \iiint \rho \log \rho \, dx_1 dx_2 dx_3 \\
&= -\iiint \rho\left(\Phi^{-1}(x_1, x_2, x_3)\right) |J^{-1}| \log\left[\rho\left(\Phi^{-1}(x_1, x_2, x_3)\right) |J^{-1}|\right] dx_1 dx_2 dx_3 \\
&\quad + \iiint \rho \log \rho \, dx_1 dx_2 dx_3 \\
&= -\iiint \rho(v_1, v_2, v_3)|J^{-1}| \left[\log \rho(v_1, v_2, v_3) + \log|J^{-1}|\right] |J| dv_1 dv_2 dv_3 \\
&\quad + \iiint \rho \log \rho \, dx_1 dx_2 dx_3 \\
&= -\iiint \rho(x_1, x_2, x_3) \log|J^{-1}| dx_1 dx_2 dx_3,
\end{aligned}$$

concisely rewritten as

$$\Delta H = E \log|J|. \tag{A2}$$

Meantime, in the case when Φ_k is invertible of 3D transformations,

$$\Delta H_k^* = E \log|J_k|. \tag{A3}$$

The entropy of X_k increases as

$$\begin{aligned}
\Delta H_k = &-\int_{\Omega_k} \left(\iint_{\Omega_i \times \Omega_j} P\rho \, dx_i dx_j\right) \log\left(\iint_{\Omega_i \times \Omega_j} P\rho \, dx_i dx_j\right) dx_k \\
&+ \int_{\Omega_k} \rho_k \log \rho_k dx_k,
\end{aligned} \tag{A4}$$

where ρ_k is the marginal density of X_k.
When Φ_k is noninvertible,

$$\begin{aligned}
\Delta H_k^* = &\int \rho_k(x_k) \log \rho_k(x_k) dx_k \\
&- \iiint P_k \rho_k \left(\Phi_k(x_i, x_j, x_k)\right) \log P_k \rho_k \left(\Phi_k(x_i, x_j, x_k)\right) \rho(x_i, x_j|x_k)|J_k| dx_i dx_j dx_k,
\end{aligned} \tag{A5}$$

where P_k is the $F - P$ operator when x_i, x_j is frozen as parameters in P_k. It is easy to find that Equation (A5) reduces to Equation (A3) when Φ_k is invertible. Therefore, the entropy transfers from X_i, X_j to X_k can be unified into a form

$$\begin{aligned}
T_{i,j \to k} = &-\int_{\Omega_k} \left(\iint_{\Omega_i \times \Omega_j} P\rho \, dx_i dx_j\right) \log\left(\iint_{\Omega_i \times \Omega_i} P\rho \, dx_i dx_j\right) dx_k \\
&+ \iiint P_k \rho_k \left(\Phi_k(x_i, x_j, x_k)\right) \log P_k \rho_k \left(\Phi_k(x_i, x_j, x_k)\right) \rho(x_i, x_j|x_k)|J_k| dx_i dx_j dx_k,
\end{aligned} \tag{A6}$$

where $i, j, k = 1, 2, 3$ with different i, j, k at the same time.

Just as the former case with continuous variables, the information flow obtained by Equation (A6) has the following property:

Theorem A1. *If Φ_k is independent of x_i, x_j in system (A1) with different i, j, k, then $T_{i,j\to k} = 0$.*

The detailed proof of Theorem A1 is presented in Appendix A.1. Moreover, the formalism of 3D system can be reduced to the formalism in 2D cases with the previously mentioned conditions being satisfied. For example, when Φ_k has no dependence on x_i,

$$
\begin{aligned}
T_{i,j\to k} =& -\int_{\Omega_k}\left(\iint_{\Omega_i\times\Omega_j} P\rho\,dx_i dx_j\right)\log\left(\iint_{\Omega_i\times\Omega_j} P\rho\,dx_i dx_j\right)dx_k \\
&+ \iiint P_k\rho_k\left(\Phi_k(x_i,x_j,x_k)\right)\log P_k\rho_k\left(\Phi_k(x_i,x_j,x_k)\right)\rho(x_i,x_j|x_k)|J_k|dx_i dx_j dx_k \\
=& -\int P_k\rho_k\log P_k\rho_k\,dx_k \\
&+ \iint P_k\rho_k(\Phi_k(x_i,x_j,x_k)\log P_k\rho_k(\Phi_i(x_i,x_j,x_k))|J_k|\left(\int_{\Omega_i}\rho(x_i,x_j|x_k)dx_i\right)dx_i dx_j \\
=& -\int P_k\rho_k\log P_k\rho_k\,dx_k \\
&+ \iint P_k\rho_k\left(\Phi_k(x_j,x_k)\right)\log P_k\rho_k\left(\Phi_k(x_j,x_k)\right)\rho(x_j|x_k)|J_k|dx_j dx_k = T_{j\to k}.
\end{aligned}
$$

In particular, when Φ_k has no dependence on x_i and x_j,

$$
T_{i,j\to k} = T_{i\to k} = T_{j\to k} = 0.
$$

Furthermore, when Φ_k has dependence on x_i and x_j,

$$
\begin{aligned}
T_{i,j\to k} =& -\int_{\Omega_k}\left(\iint_{\Omega_i\times\Omega_j} P\rho\,dx_i dx_j\right)\log\left(\iint_{\Omega_i\times\Omega_j} P\rho\,dx_i dx_j\right)dx_k \\
&+ \iiint P_k\rho_k\left(\Phi_k(x_i,x_j,x_k)\right)\log P_k\rho_k\left(\Phi_k(x_i,x_j,x_k)\right)\rho(x_i,x_j|x_k)|J_k|dx_i dx_j dx_k \\
=& -\int_{\Omega_k}\left(\iint_{\Omega_i\times\Omega_j} P\rho\,dx_i dx_j\right)\log\left(\iint_{\Omega_i\times\Omega_j} P\rho\,dx_i dx_j\right)dx_k \\
&+ \iiint P_k\rho_k\left(\Phi_k(x_i,x_j,x_k)\right)\log P_k\rho_k\left(\Phi_k(x_i,x_j,x_k)\right)\rho(x_j|x_k)\rho(x_i|x_j,x_k)|J_k|dx_i dx_j dx_k \\
=& -\int_{\Omega_k}\left(\iint_{\Omega_i\times\Omega_j} P\rho\,dx_i dx_j\right)\log\left(\iint_{\Omega_i\times\Omega_j} P\rho\,dx_i dx_j\right)dx_k \\
&+ \int_{\Omega_i}\left(\iint_{\Omega_j\times\Omega_k} P_k\rho_k\left(\Phi_k(x_i,x_j,x_k)\right)\log P_k\rho_k\left(\Phi_k(x_i,x_j,x_k)\right)\rho(x_j|x_k)|J_k|dx_j dx_k\right) \\
&\cdot \rho(x_i|x_j,x_k)dx_i \\
=& -\int_{\Omega_i} T_{j\to k}\cdot\rho(x_i|x_j,x_k)dx_i \\
&+ \int_{\Omega_i}\left(\int_{\Omega_k} P_k\rho_k\log P_k\rho_k\,dx_k\right)\cdot\rho(x_i|x_j,x_k)dx_i - \int_{\Omega_k} P_k\rho_k\log P_k\rho_k\,dx_k
\end{aligned}
$$

The above formalisms can also be generalized to n-dimensional systems by efficient processing of the relationship between the $F - P$ operator

$$
\int_w P\rho(x_1,x_2,\ldots x_n)dx_1 dx_2\ldots dx_n = \int_{\Phi^{-1}(w)}\rho(x_1,x_2,\ldots x_n)dx_1 dx_2\ldots dx_n
$$

and the entropy evolution at different time steps. For example, the transfer of entropy from $X_2, X_3, \ldots, X_n$ to X_1 is

$$
\begin{aligned}
T_{2,3,\ldots,n \to 1} = {}& -\int_{\Omega_1} \left(\int_{\Omega_{2\ldots n}} P\rho\, dx_2 dx_3 \ldots dx_n \right) \log \left(\int_{\Omega_{2\ldots n}} P\rho\, dx_2 dx_3 \ldots dx_n \right) dx_1 \\
& + \int_{\Omega} P_1\rho_1 \left(\Phi_1(x_1, x_2, \ldots, x_n) \right) \log P_1\rho_1 \left(\Phi_1(x_1, x_2, \ldots, x_n) \right) \\
& \cdot \rho(x_2, x_3, \ldots, x_n | x_1) |J_1| dx_1 dx_2 \ldots dx_n.
\end{aligned}
$$

Here $\Omega_{2\ldots n}$ is the simplified script of $\Omega_2 \times \Omega_3 \times \ldots \times \Omega_n$. Similar to the continuous cases, the generalized version of the property of Theorem A1 is also suitable for multi-dimensional discrete mappings.

Appendix A.1.

Proof of Theorem A1. We only need to show that when Φ_k is independent of x_i, x_j in 3D system,

$$
\Delta H_k^* = \Delta H_k.
$$

According to Equation (A4) and Equation (A5), we only need to prove

$$
\begin{aligned}
& -\iiint_{\Omega} P_k\rho_k \left(\Phi_k(x_k, x_i, x_j) \right) \log P_k\rho_k \left(\Phi_k(x_k, x_i, x_j) \right) \rho(x_i, x_j | x_k) |J_k| dx_i dx_j dx_k \\
={}& -\int_{\Omega_k} \left(\iint_{\Omega_i \times \Omega_j} P\rho\, dx_i dx_j \right) \log \left(\iint_{\Omega_i \times \Omega_j} P\rho\, dx_i dx_j \right) dx_k.
\end{aligned}
$$

According to the definition of the $F - P$ operator and the condition that Φ_k is independent of x_i, x_j at the same time,

$$
\begin{aligned}
& -\iiint_{\Omega} P_k\rho_k \left(\Phi_k(x_k, x_i, x_j) \right) \log P_k\rho_k \left(\Phi_k(x_k, x_i, x_j) \right) \rho(x_i, x_j | x_k) |J_k| dx_i dx_j dx_k \\
={}& -\int_{\Omega_k} P_k\rho_k \left(\Phi_k(x_k, x_i, x_j) \right) \log P_k\rho_k \left(\Phi_k(x_k, x_i, x_j) \right) |J_k| \iint_{\Omega_i \times \Omega_j} \rho(x_i, x_j | x_k) dx_i dx_j dx_k
\end{aligned}
$$

because $\iint_{\Omega_i \times \Omega_j} \rho(x_i, x_j | x_k) dx_i dx_j dx_k = 1$

$$
\begin{aligned}
& -\int_{\Omega_k} P_k\rho_k \left(\Phi_k(x_k, x_i, x_j) \right) \log P_k\rho_k \left(\Phi_k(x_k, x_i, x_j) \right) |J_k| \iint_{\Omega_i \times \Omega_j} \rho(x_i, x_j | x_k) dx_i dx_j dx_k \\
={}& -\int_{\Omega_k} P_k\rho_k(y_k) \log P_k\rho_k(y_k) dy_k \\
={}& -\int_{\Omega_k} p_k\rho_k(x_k) \log P_k\rho_k(x_k) dx_k \\
={}& -\int_{\Omega_k} \left(\iint_{\Omega_i \times \Omega_j} P\rho\, dx_i dx_j \right) \log \left(\iint_{\Omega_i \times \Omega_j} P\rho\, dx_i dx_j \right) dx_k
\end{aligned}
$$

where $y_1 = \Phi_k(x_k, x_i, x_j)$. So

$$
\begin{aligned}
\Delta H_k^* &= \int \rho_k(x_k) \log \rho_k(x_k) dx_k \\
&\quad - \iiint P_k \rho_k \left(\Phi_k(x_k, x_i, x_j) \right) \log P_k \rho_k \left(\Phi_k(x_k, x_i, x_j) \right) \rho(x_i, x_j | x_k) |J_k| dx_i dx_j dx_k \\
&= - \int_{\Omega_k} \left(\iint_{\Omega_i \times \Omega_j} P \rho dx_i dx_j \right) \log \left(\iint_{\Omega_i \times \Omega_j} P \rho dx_i dx_j \right) dx_k \\
&\quad + \int_{\Omega_k} \rho_k \log \rho_k dx_k = \Delta H_k.
\end{aligned}
$$

$\square$

References

1. Schreiber, T. Measuring Information Transfer. *Phys. Rev. Lett.* **2000**, *85*, 461. [CrossRef] [PubMed]
2. Horowitz, J.M.; Esposito, M. Thermodynamics with Continuous Information Flow. *Phys. Rev. X* **2014**, *4*, 031015. [CrossRef]
3. Cafaro, C.; Ali, S.A.; Giffin, A. Thermodynamic aspects of information transfer in complex dynamical systems. *Phys. Rev. E* **2016**, *93*, 022114. [CrossRef] [PubMed]
4. Gencaga, D.; Knuth, K.H.; Rossow, W.B. A Recipe for the Estimation of Information Flow in a Dynamical System. *Entropy* **2015**, *17*, 438–470. [CrossRef]
5. Kwapien, J.; Drozdz, S. Physical approach to complex systems. *Phys. Rep.* **2012**, *515*, 115–226. [CrossRef]
6. Liang, X.S.; Kleeman, R. Information Transfer between Dynamical Systems Components. *Phys. Rev. Lett.* **2005**, *95*, 244101. [CrossRef] [PubMed]
7. Kleeman, R. Information flow in ensemble weather predictions. *J. Atmos. Sci* **2007**, *6*, 1005–1016. [CrossRef]
8. Touchette, H.; Lloyd, S. Information-Theoretic Limits of Control. *Phys. Rev. Lett.* **2000**, *84*, 1156. [CrossRef] [PubMed]
9. Touchette, H.; Lloyd, S. Information-theoretic approach to the study of control systems. *Phys. A* **2004**, *331*, 140. [CrossRef]
10. Sun, J.; Cafaro, C.; Bollt, E.M. Identifying coupling structure in complex systems through the optimal causation entropy principle. *Entropy* **2014**, *16*, 3416–3433. [CrossRef]
11. Cafaro, C.; Lord, W.M.; Sun, J.; Bollt, E.M. Causation entropy from symbolic representations of dynamical systems. *CHAOS* **2015**, *25*, 043106. [CrossRef] [PubMed]
12. Majda, A.; Kleeman, R.; Cai, D. A Framework for Predictability through Relative Entropy. *Methods Appl. Anal.* **2002**, *9*, 425–444.
13. Haven, K.; Majda, A.; Abramov, R. Quantifying predictability through information theory: Small-sample estimation in a non-Gaussian framework. *J. Comp. Phys.* **2005**, *206*, 334–362. [CrossRef]
14. Abramov, R.V.; Majda, A.J. Quantifying Uncertainty for Non-Gaussian Ensembles in Complex Systems. *SIAM J. Sci. Stat. Comp.* **2004**, *26*, 411–447. [CrossRef]
15. Kaiser, A.; Schreiber, T. Information transfer in continuous processes. *Phys. D* **2002**, *166*, 43–62. [CrossRef]
16. Abarbanel, H.D.I.; Masuda, N.; Rabinovich, M.I.; Tumer, E. Distribution of Mutual Information. *Phys. Lett. A* **2001**, *281*, 368–373. [CrossRef]
17. Wyner, A.D.; Mackey, M.C. A definition of conditional mutual information for arbitrary ensembles. *Inf. Control* **1978**, *38*, 51–59. [CrossRef]
18. Lasota, A.; Mackey, M.C. *Chaos, Fractals, and Noise: Stochastic Aspects of Dynamics*; Springer: New York, NY, USA, 1994.
19. Liang, X.S.; Kleeman, R. A rigorous formalism of information transfer between dynamical system components. I. Discrete mapping. *Physica D* **2007**, *231*, 1–9. [CrossRef]
20. Liang, X.S.; Kleeman, R. A rigorous formalism of information transfer between dynamical system components. II. Continuous flow. *Physica D* **2007**, *227*, 173–182. [CrossRef]
21. Liang, X.S. Information flow within stochastic dynamical systems. *Phys. Rev. E* **2008**, *78*, 031113. [CrossRef] [PubMed]

22. Liang, X.S. Uncertainty generation in deterministic fluid flows: Theory and applications with an atmospheric stability model. *Dyn. Atmos. Oceans* **2011**, *52*, 51–79. [CrossRef]
23. Liang, X.S. The Liang-Kleeman information flow: Theory and application. *Entropy* **2013**, *15*, 327–360. [CrossRef]
24. Liang, X.S. Unraveling the cause-effect relation between time series. *Phys. Rev. E* **2014**, *90*, 052150. [CrossRef] [PubMed]
25. Liang, X.S. Information flow and causality as rigorous notions ab initio. *Phys. Rev. E* **2016**, *94*, 052201. [CrossRef] [PubMed]
26. Majda, A.J.; Harlim, J. Information flow between subspaces of complex dynamical systems. *Proc. Natl. Acad. Sci. USA* **2007**, *104*, 9558–9563. [CrossRef]
27. Zhao, X.J.; Shang, P.J. Measuring the uncertainty of coupling. *Europhys. Lett.* **2015**, *110*, 60007. [CrossRef]
28. Zhao, X.J.; Shang, P.J.; Huang, J.J. Permutation complexity and dependence measures of time series. *Europhys. Lett.* **2013**, *102*, 40005. [CrossRef]
29. Iooss, B.; Lemaitre, P. A review on global sensitivity analysis methods. *Oper. Res. Comput. Sci. Interfaces* **2014**, *59*, 101–122.
30. Borgonovo, E.; Plischke, E. Sensitivity analysis: A review of recent advances. *Eur. J. Oper. Res.* **2016**, *248*, 869–887. [CrossRef]
31. Auder, B.; Crecy, A.D.; Iooss, B.; Marques, M. Screening and metamodeling of computer experiments with functional outputs. Application to thermal–hydraulic computations. *Reliab. Eng. Syst. Safety* **2012**, *107*, 122–131. [CrossRef]
32. Zhao, X.J.; Shang, P.J.; Lin, A.J. Transfer mutual information: A new method for measuring information transfer to the interactions of time series. *Physica A* **2017**, *467*, 517–526. [CrossRef]
33. Lorenz, E.N. Deterministic nonperiodic flow. *J. Atmos. Sci.* **1963**, *20*, 130–141. [CrossRef]
34. Nepomuceno, E.G.; Mendes, E.M.A.M. On the analysis of pseudo-orbits of continuous chaotic nonlinear systems simulated using discretization schemes in a digital computer. *Chaos Soliton Fract.* **2017**, *95*, 21–32. [CrossRef]
35. Brunton, S.L.; Brunton, B.W.; Proctor, J.L.; Kaiser, E.; Kutz, J.N. Chaos as an intermittently forced linear system. *Nat. Commun.* **2017**. [CrossRef] [PubMed]
36. Campbell, K.; McKay, M.D.; Williams, B.J. Sensitivity analysis when model outputs are functions. *Reliab. Eng. Syst. Saf.* **2006**, *91*, 1468–1472. [CrossRef]
37. Lamboni, M.; Monod, H.; Makowski, D. Multivariate sensitivity analysis to measure global contribution of input factors in dynamic models. *Reliab. Eng. Syst. Saf.* **2011**, *96*, 450–459. [CrossRef]
38. Marrel, A.; Iooss, B.; Jullien, M.; Laurent, B.; Volkova, E. Global sensitivity analysis for models with spatially dependent outputs. *Environmetrics* **2011**, *22*, 383–397. [CrossRef]
39. Loonen, R.C.G.M.; Hensen, J.L.M. Dynamic sensitivity analysis for performance-based building design and operation. In Proceedings of the BS2013: 13th Conference of International Building Performance Simulation Association, Chambéry, France, 26– 28 August 2013; pp. 299–305.
40. Richard, R.; Casas, J.; McCauley, E. Sensitivity analysis of continuous-time models for ecological and evolutionary theories. *Theor. Ecol.* **2015**. [CrossRef]
41. Chua, L.O.; Komuro, M.; Matsumoto, T. The double scroll family: Parts I and II. *IEEE Trans. Circuits Syst.* **1986**, *CAS-33(11)*, 1072–1118. [CrossRef]
42. Chua, L.O. The genesis of Chua's circuit. In *Archiv fur Elektronik und Ubertragungstechnik*; University of California: Berkeley, CA, USA, 1992; Volume 46, pp. 250–257.
43. Chua, L.O. A zoo of strange attractor from the canonical Chua's circuits. In Proceedings of the 35th Midwest Symposium on Circuits and Systems, Washington, DC, USA, 1992; pp. 916–926.
44. Liao, X.X.; Yu, P.; Xie, S.L.; Fu, Y.L. Study on the global property of the smooth Chua's system. *Int. J. Bifurcat. Chaos Appl. Sci. Eng.* **2006**, *16*, 2815–2841. [CrossRef]
45. Zhou, G.P.; Huang, J.H.; Liao, X.X.; Cheng, S.J. Stability Analysis and Control of a New Smooth Chua's System. *Abstract Appl. Anal.* **2013**, *2013*, 10 pages, http://dx.doi.org/10.1155/2013/620286. [CrossRef]
46. Bertacchini, F.; Bilotta, E.; Gabriele, L.; Pantano, P.; Tavernise, A. Toward the Use of Chua's Circuit in Education, Art and Interdisciplinary Research: Some Implementation and Opportunities. *LEONARDO* **2013**, *46*, 456–463. [CrossRef]

47. Bilotta, E.; Blasi, G.D.; Stranges, F.; Pantano, P. A Gallery of Chua Attractors. Part V. *Int. J. Bifurcat. Chaos* **2007**, *17*, 1383–1511. [CrossRef]
48. Adamo, A.; Tavernise, A. Generation of Ego dynamics. In Proceedings of the VIII International Conference on Generative Art, Milan, Italy, 15–17 December 2007.
49. Kingni, S.T.; Jafari, S.; Simo, H.; Woafo, P. Three-dimensional chaotic autonomous system with only one stable equilibrium: Analysis, circuit design, parameter estimation, control, synchronization and its fractional-order form. *Eur. Phys. J. Plus* **2014**, *129*. [CrossRef]

Article

Fractional Form of a Chaotic Map without Fixed Points: Chaos, Entropy and Control

**Adel Ouannas [1], Xiong Wang [2], Amina-Aicha Khennaoui [3], Samir Bendoukha [4], Viet-Thanh Pham [5,* and Fawaz E. Alsaadi [6]

[1] Department of Mathematics and Computer Science, University of Larbi Tebessi, Tebessa 12002, Algeria; Ouannas@mail.univ-tebessa.dz

[2] Institute for Advanced Study, Shenzhen University, Shenzhen 518060, Guangdong, China; wangxiong8686@szu.edu.cn

[3] Department of Mathematics and Computer Sciences, University of Larbi Ben M'hidi, Oum El Bouaghi 04000, Algeria; Khennaoui@mail.univ-tebessa.dz

[4] Electrical Engineering Department, College of Engineering at Yanbu, Taibah University, Medina 42353, Saudi Arabia; sbendoukha@taibahu.edu.sa

[5] Modeling Evolutionary Algorithms Simulation and Artificial Intelligence, Faculty of Electrical & Electronics Engineering, Ton Duc Thang University, Ho Chi Minh City, Vietnam

[6] Department of Information Technology, Faculty of Computing and IT, King Abdulaziz University, Jeddah 21589, Saudi Arabia; fawazkau@gmail.com

* Correspondence: phamviethanh@tdt.edu.vn

Received: 1 August 2018; Accepted: 15 September 2018; Published: 20 September 2018

Abstract: In this paper, we investigate the dynamics of a fractional order chaotic map corresponding to a recently developed standard map that exhibits a chaotic behavior with no fixed point. This is the first study to explore a fractional chaotic map without a fixed point. In our investigation, we use phase plots and bifurcation diagrams to examine the dynamics of the fractional map and assess the effect of varying the fractional order. We also use the approximate entropy measure to quantify the level of chaos in the fractional map. In addition, we propose a one-dimensional stabilization controller and establish its asymptotic convergence by means of the linearization method.

Keywords: discrete chaos; discrete fractional calculus; hidden attractors; approximate entropy; stabilization

---

1. Introduction

Over the last few decades and since the Hénon map was first proposed [1], discrete-time chaotic dynamical systems have received a great deal of attention from numerous disciplines due to their ability to model various natural phenomena [2]. Throughout the years, a variety of chaotic maps has been proposed and their dynamics investigated [3–7]. The dynamics and control of such systems have been widely investigated [8–13]. Recently, an interesting investigation was carried out in [14] on a rather general 2D map that can, under certain parameters, have no fixed points and possess hidden dynamics. The authors examined the stability of the fixed points and showed that the map exhibits rich dynamics and may in some instances have a very tiny basin of attraction. Note that the term hidden attractors first came about in the investigation of continuous chaotic systems [15–17]. They refer to attractors that do not contain the neighborhoods of the equilibria. The hidden attractor property plays important roles in science and engineering [18–21]. The existence of such attractors in many engineering applications is considered problematic and requires stabilization [22–24]. The authors considered a new system inspired by the logistic map and examined its bifurcation and hidden dynamics. The schematic approach proposed in [14] for studying such hidden dynamics has been picked up by researchers such as [25].

In recent years, with the growing advancement in the field of discrete fractional calculus, a few studies have emerged considering the dynamics, control and applications of fractional chaotic maps [26–32]. In addition, there are few works related to chaotic maps with hidden attractors [14,25]. Especially, research to date has not yet studied fractional chaotic maps without a fixed point. It should be noted that although the inception of fractional continuous calculus took place centuries ago, its discrete counterpart was not properly explored until recently [33]. The first definition of a fractional difference operator was made by Diaz and Olser in 1974 [34]. In fact, the vast majority of available literature on the subject was published in the last decade, including [35–40].

In this paper, we examine the dynamics the fractional version of the general map proposed in [14] by means of phase plots and bifurcation diagrams. It is noted that there is no fixed point in such a chaotic fractional map. There is no fractional-order chaotic map without fixed points reported in the literature. As far as we aware, this is the first time that a fractional-order chaotic map without fixed points has been investigated. We show that the fractional order has a major impact on the chaotic range and the shape of hidden attractors. We also use the approximate entropy measure to quantify the level of chaotic behavior present in the fractional map. By varying the fractional order, we show that it has an impact on the entropy. We also propose a one-dimensional stabilization controller that forces the system states to zero asymptotically. Throughout our analysis, we make use of numerical methods to confirm the findings.

2. The Fractional Map without Fixed Points

In this paper, we are interested in the dynamics, entropy and control of the fractional map based on the standard iterated map of the form:

$$\begin{cases} x(n+1) = y(n), \\ y(n+1) = x(n) + a_1 x^2(n) + a_2 y^2(n) - a_3 x(n) y(n) - a_4, \end{cases} \tag{1}$$

where a_1, a_2, a_3 and a_4 are some real-value parameters. This map was developed by Jiang et al. [14] as a variation of the original Hénon map [1]. They showed by means of analytical and numerical methods that hidden chaotic attractors exist in the map for certain values of the parameters a_1, a_2, a_3 and a_4. The phase-space portraits of the map with no fixed point are depicted in Figure 1 for three typical examples. This map has some interesting dynamics with hidden strange attractors. The authors showed that chaos exists in the map with different scenarios: no fixed point, a single fixed point and two fixed points. In the following, we develop a fractional chaotic map based on (1) to examine its dynamics and control.

Before we can state the fractional map we are concerned with, let us recall some important aspects of discrete fractional calculus. First of all, consider a generic function $X(t) : \mathbb{N}_a \rightarrow \mathbb{R}$ where $\mathbb{N}_a$ denotes the set of all discrete numbers starting from a, i.e., $\mathbb{N}_a = \{a, a+1, a+2, ...\}$. Given a fractional difference number $v > 0$ and the function $\sigma(s) = s + 1$, we define the v-th fractional sum of $X(t)$ similar to [35] as:

$$\Delta_a^{-v} X(t) = \frac{1}{\Gamma(v)} \sum_{s=a}^{t-v} (t - \sigma(s))^{(v-1)} X(s), \tag{2}$$

for all $t \in \mathbb{N}_{a+n-v}$ and with t^v being the falling function defined in terms of the Gamma function Γ as:

$$t^v = \frac{\Gamma(t+1)}{\Gamma(t+1-v)}. \tag{3}$$

With this in mind, we may define the v-th Caputo type delta difference of $X(t)$ similar to [36] by:

$$^C\Delta_a^v X(t) = \Delta_a^{-(n-v)} \Delta^n X(t) = \frac{1}{\Gamma(n-v)} \sum_{s=a}^{t-(n-v)} (t - \sigma(s))^{(n-v-1)} \Delta_s^n X(s), \tag{4}$$

where $v \notin \mathbb{N}$ is the fractional order, $t \in \mathbb{N}_{a+n-v}$ and $n = [v] + 1$.

Now, that we have stated the basics of discrete fractional calculus, we may start our analysis. System (1) can be rewritten in difference form as:

$$\begin{cases} \Delta x\,(n) = y\,(n) - x\,(n)\,, \\ \Delta y\,(n) = x\,(n) + a_1 x^2\,(n) + a_2 y^2\,(n) - a_3 x\,(n)\,y\,(n) - a_4 - y\,(n)\,. \end{cases} \tag{5}$$

Then, using the Caputo difference operator ${}^C\Delta_a^v$ as defined in (4), we obtain the fractional version of the map for $t \in \mathbb{N}_{a+1-v}$ and $0 < v \leq 1$ as:

$$\begin{cases} {}^C\Delta_a^v x\,(t) = y\,(t-1+v) - x\,(t-1+v)\,, \\ {}^C\Delta_a^v y\,(t) = x\,(t-1+v) + a_1 x^2\,(t-1+v) + a_2 y^2\,(t-1+v) \\ \qquad\qquad -a_3 x\,(t-1+v)\,y\,(t-1+v) - a_4 - y\,(t-1+v)\,. \end{cases} \tag{6}$$

From here on, we will refer to (6) as the fractional map. Our new fractional-order map belongs to a special class of dynamical systems with "hidden attractors", which have received significant attention recently [19,24]. We believe that our work will assist researchers in further understanding systems with hidden attractors.

In order to examine the dynamics of the fractional map (6), we must develop a numerical formula for it. Let us recall an important theorem that defines the equivalent discrete integral equation corresponding to a generic fractional difference, which will enable us to obtain our numerical formula.

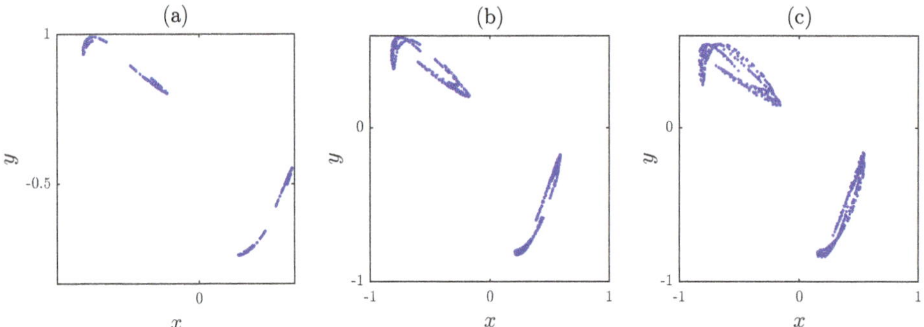

Figure 1. Phase plots of the map without a fixed point for: (a) $(a_1, a_2, a_3, a_4) = (0.2, 0.71, 0.91, 1.14)$ and $(x(0), y(0)) = (0.93, -0.44)$; (b) $(a_1, a_2, a_3, a_4) = (0.51, 1, 1.51, 0.74)$ and $(x(0), y(0)) = (-0.81, 0.51)$; (c) $(a_1, a_2, a_3, a_4) = (0.6, 1, 1.6, 0.72)$ and $(x(0), y(0)) = (-0.26, 0.18)$.

Theorem 1 ([41])**.** *For the delta fractional difference equation:*

$$\begin{cases} {}^C\Delta_a^v u\,(t) = f\,(t+v-1, u\,(t+v-1))\,, \\ \Delta^k = u_k,\ n = [v] + 1,\ k = 0, 1, ..., n-1, \end{cases} \tag{7}$$

the equivalent discrete integral equation can be obtained as:

$$u\,(t) = u_0\,(t) + \frac{1}{\Gamma\,(v)} \sum_{s=a+n-v}^{t-v} (t - \sigma\,(s))^{(v-1)} f\,(s+v-1, u\,(s+v-1))\,,\ t \in \mathbb{N}_{\alpha+n}, \tag{8}$$

where:

$$u_0\,(t) = \sum_{k=0}^{m-1} \frac{(t-a)^k}{k} \Delta^k u\,(a)\,. \tag{9}$$

By applying Theorem 1, we can state the equivalent discrete integral form of (6) for $t \in \mathbb{N}_{a+1}$ as:

$$\begin{cases} x(t) = x(a) + \frac{1}{\Gamma(v)} \sum_{s=a+1-v}^{t-v} (t - \sigma(s))^{(v-1)} (y(s+v-1) - x(s+v-1)), \\ y(t) = y(a) + \frac{1}{\Gamma(v)} \sum_{s=a+1-v}^{t-v} (t - \sigma(s))^{(v-1)} \left(x(s+v-1) + a_1 x^2(s+v-1) \right. \\ \qquad \left. + a_2 y^2(s+v-1) - a_3 x(s+v-1) y(s+v-1) - a_4 - y(s+v-1) \right). \end{cases} \tag{10}$$

The reciprocal $\frac{(t-\sigma(s))^{(v-1)}}{\Gamma(v)}$ is known as a discrete kernel function. For simplicity, we may choose:

$$\frac{(t - \sigma(s))^{(v-1)}}{\Gamma(v)} = \frac{\Gamma(t-s)}{\Gamma(v)\Gamma(t-s-v+1)}. \tag{11}$$

This leads to the following numerical formulas for $a = 0$:

$$\begin{cases} x(n) = x(0) + \frac{1}{\Gamma(v)} \sum_{j=1}^{n} \frac{\Gamma(n-j+v)}{\Gamma(n-j+1)} (y(j-1) - x(j-1)), \\ y(n) = y(0) + \frac{1}{\Gamma(v)} \sum_{j=1}^{n} \frac{\Gamma(n-j+v)}{\Gamma(n-j+1)} \left(x(j-1) + a_1 x^2(j-1) \right. \\ \qquad \left. + a_2 y^2(j-1) - a_3 x(j-1) y(j-1) - a_4 - y(j-1) \right), \end{cases} \tag{12}$$

These numerical formulas will allow us to plot phase-space portraits, bifurcation diagrams and error convergence plots throughout the remainder of this paper.

3. Chaotic Dynamics and Entropy Analysis

3.1. Chaotic Dynamics

Now that we have our fractional map (6) and the corresponding numerical formulas (12), let us study the map's dynamics and chaotic behavior. First, we study the effect of the fractional order v on the dynamics of the map for parameter values $(a_1, a_2, a_3, a_4) = (0.2, 0.71, 0.91, 1.14)$. Evaluating (12) for $v = 1$ and with some direct calculations, we can see that the resulting dynamics of the fractional map are identical to those of the classical one even though it has a discrete memory effect, i.e., the solution $x(n)$ depends on all previous values $x(0), x(1), ..., x(n-1)$. With initial values $(x(0), y(0)) = (0.93, -0.44)$, Figure 2 shows the phase portrait of the fractional map for various values of the fractional order v. We notice that as v decreases, the trajectory $(x(t), y(t))$ remains bounded, whereas when $v \leq 0.976$, the chaotic behavior is delayed and the states of the fractional map diverge to infinity.

Next, we set the parameters (a_1, a_2, a_3, a_4)–$(0.51, 1, 1.51, 0.74)$ and choose the initial values $(x(0), y(0)) = (-0.81, 0.51)$. Figure 3 depicts the phase portraits of the fractional map for the three different fractional orders $v = 1$, $v = 0.979$ and $v = 0.963$. Similarly, when $v = 1$, the fractional map refers to the classical system. While $0.963 \leq v \leq 1$, the fractional map (6) exhibits a chaotic behavior, and when $v = 0.962$, we fall into an unbounded attractor.

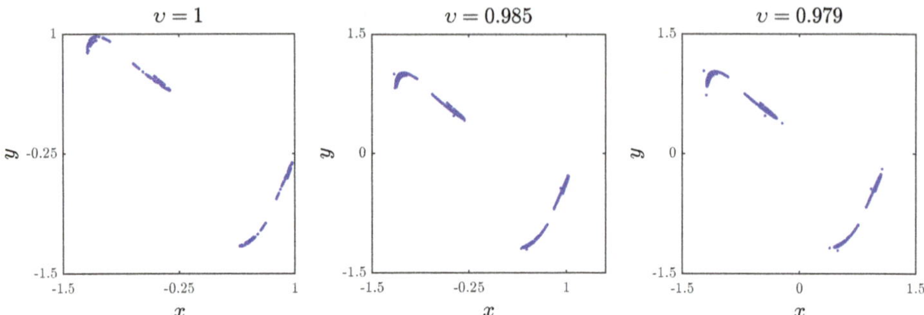

Figure 2. The chaotic attractor obtained with $(a_1, a_2, a_3, a_4) = (0.2, 0.71, 0.91, 1.14)$ and $(x(0), y(0)) = (0.93, -0.44)$ for different fractional orders v.

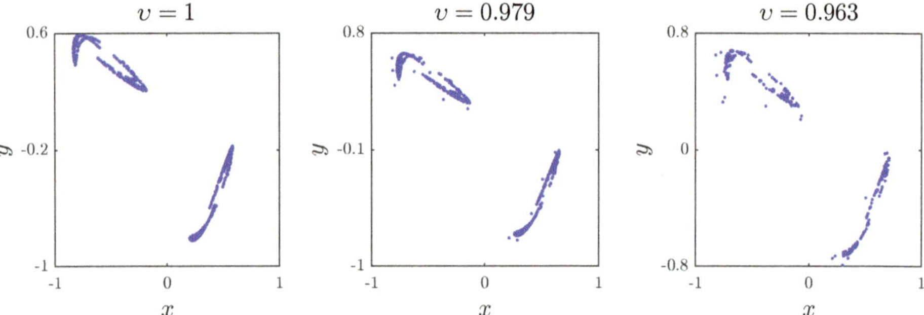

Figure 3. The chaotic attractor obtained with $(a_1, a_2, a_3, a_4) = (0.51, 1, 1.51, 0.74)$ and $(x(0), y(0)) = (-0.81, 0.51)$ for different fractional orders v.

Let us, now, consider the third set of parameters $(a_1, a_2, a_3, a_4) = (0.6, 1, 1.6, 0.72)$ with initial states $(x(0), y(0)) = (-0.26, 0.18)$. As can be seen in Figure 4, the resulting trajectories of the fractional map vary with v. When $v < 0.95$, chaos disappears completely.

The bifurcation diagrams for different parameters (a_1, a_2, a_3, a_4) are shown in Figures 5–7, respectively. First we fix parameters (a_1, a_3, a_4)–$(0.2, 0.91, 1.14)$ and vary a_2 along the interval $[0.46, 0.75]$. Clearly, decreasing the fractional order v affects the interval over which chaos is exhibited. In Figure 6, the bifurcation diagram is obtained with $(a_2, a_3, a_4) = (1, 1.51, 0.74)$ and the critical parameter a_1 being varied in steps of $\Delta a_1 = 0.0006$. In this case, when we decrease the fractional order v, the opposite is observed as the chaotic band expands and the eight-period stage disappears. Finally, Figure 7 is obtained for $(a_2, a_3, a_4) = (1, 1.6, 0.72)$ with a_1 as the critical parameter. We see that a slight change in the fractional order has a considerable effect on the dynamics of the fractional map. For completeness, the time evolution of the states belonging to the fractional map are displayed in Figure 8 for $v = 0.979$.

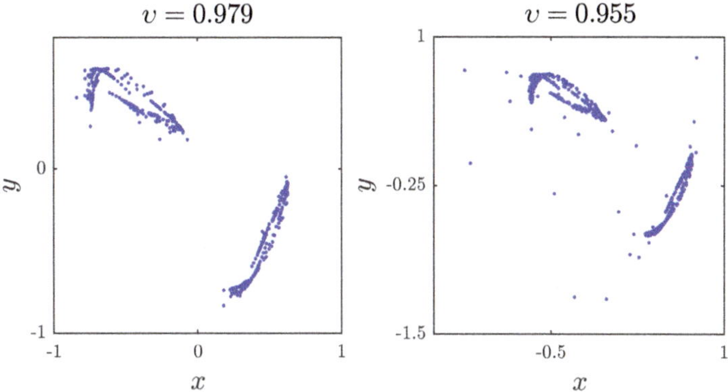

Figure 4. The chaotic attractor obtained with $(a_1, a_2, a_3, a_4) = (0.6, 1, 1.6, 0.72)$ and $(x(0), y(0)) = (-0.26, 0.18)$ for different fractional orders v.

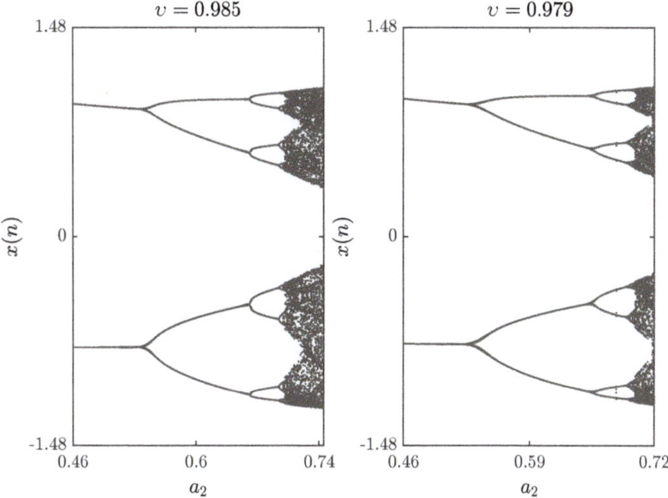

Figure 5. Bifurcation diagrams with a_2 as the critical parameter and $(a_1, a_3, a_4) = (0.2, 0.91, 1.14)$ and $(x(0), y(0)) = (0.93, -0.44)$ for different fractional orders v.

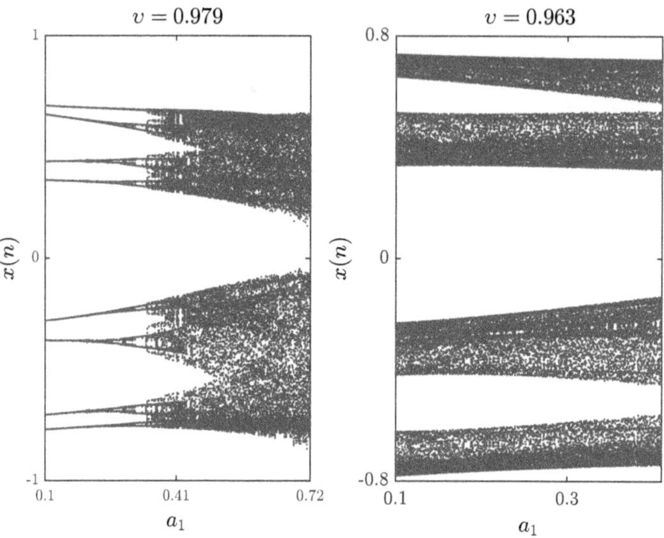

Figure 6. Bifurcation diagrams with a_1 as the critical parameter and $(a_2, a_3, a_4) = (1, 1.51, 0.74)$ and $(x(0), y(0)) = (-0.81, 0.51)$ for different fractional orders v.

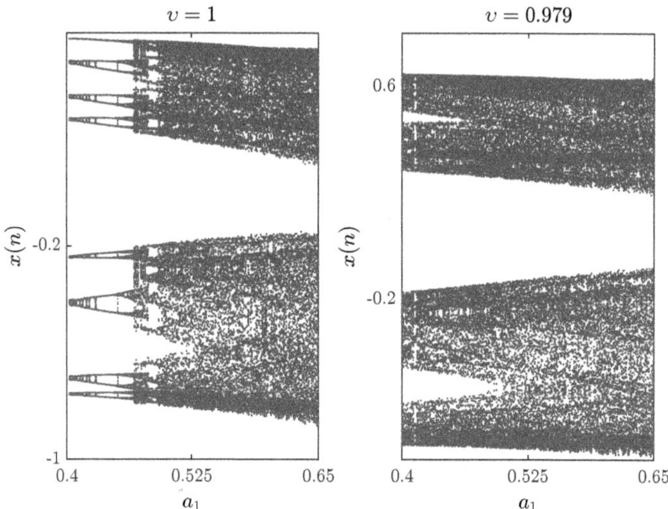

Figure 7. Bifurcation diagrams with a_1 as the critical parameter and $(a_2, a_3, a_4) = (1, 1.6, 0.72)$ and $(x(0), y(0)) = (-0.26, 0.18)$ for different fractional orders v.

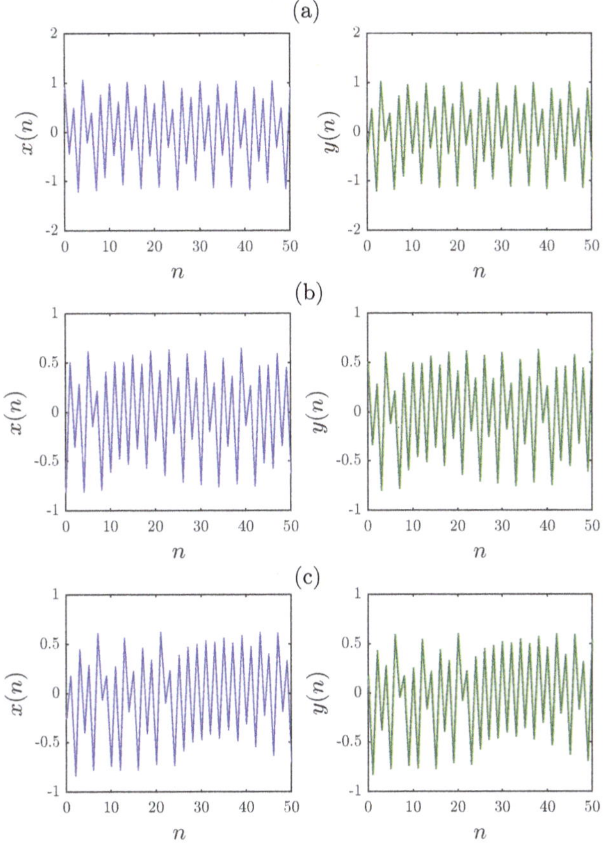

Figure 8. Time evolution of states for $v = 0.979$ and: (**a**) $(a_1, a_2, a_3, a_4) = (0.2, 0.71, 0.91, 1.14)$ and $(x(0), y(0)) = (0.93, -0.44)$; (**b**) $(a_1, a_2, a_3, a_4) = (0.51, 1, 1.51, 0.74)$ and $(x(0), y(0)) = (-0.81, 0.51)$; (**c**) $(a_1, a_2, a_3, a_4) = (0.6, 1, 1.6, 0.72)$ and $(x(0), y(0)) = (-0.26, 0.18)$.

3.2. Entropy Analysis

In information theory, entropy is a logarithmic measure that quantifies the rate of transfer or generation of information in a particular system. For discrete-time dynamical systems in general, Kolmogorov–Sinai (KS) entropy is an interesting measure. A direct time-series approximation of the KS entropy was developed in [42], termed Eckmann–Ruelle (ER) entropy, which quickly became appealing as a way of quantifying the level of chaos present in a particular system. The idea is that instead of looking at the phase plots of bifurcation diagrams, an exact measure of the information generated in a sequence is more indicative of the level of chaos. Calculating the exact ER entropy experimentally is impossible. Rather, an approximate entropy (ApEn) measure was proposed in [43,44]. ApEn has been used extensively in the literature to investigate chaos in discrete dynamical systems [25,45]. In order to examine the level of chaotic behavior present in the fractional map (6), we have measured its approximate entropy (ApEn) and listed it in Table 1. We have calculated the approximate entropy using the known algorithm [43,44]. As can be seen from Table 1, the results match with the ones shown in Figures 2 and 3. For the set of parameters $(a_1, a_2, a_3, a_4) = (0.2, 0.71, 0.91, 1.14)$ and $(x(0), y(0)) = (0.93, -0.44)$, the complexity of the fractional map without a fixed point is increased when reducing the value of different fractional order v. The complexity of the fractional map without a fixed point changes when decreasing the value of v for $(a_1, a_2, a_3, a_4) = (0.51, 1, 1.51, 0.74)$ and $(x(0), y(0)) = (-0.81, 0.51)$.

Table 1. Approximate entropy calculation of the fractional map for different values of parameters and v.

Case	v	ApEn
Figure 2	1	0.0903
Figure 2	0.985	0.0955
Figure 2	0.979	0.1094
Figure 3	1	0.2083
Figure 3	0.979	0.1766
Figure 3	0.963	0.2184

4. Stabilization Control

In this section, we aim to propose a one-dimensional control law that stabilizes the states of our fractional map (6). Stabilization refers to the adaptive control of one or more system states to ensure all of the states converge asymptotically towards an equilibrium point. In our case, we assume the equilibrium to be the all zero state. Before we present our result, we recall an important theorem related to the asymptotic stability of discrete fractional systems through system linearization.

Theorem 2 ([37]). *Given a vector-valued function* $X(t) = (x_1(t), ..., x_n(t))^T$, $0 < v \le 1$, $A \in R^{n \times n}$ *and* $\forall t \in \mathbb{N}_{a+1-v}$, *the zero equilibrium of system:*

$$^C\Delta_a^v X(t) = AX(t + v - 1),\tag{13}$$

is asymptotically stable if:

$$\lambda \in \left\{ z \in \mathbb{C} : |z| < \left(2\cos \frac{|\arg z| - \pi}{2 - v} \right)^v \text{ and } |\arg z| > \frac{v\pi}{2} \right\},\tag{14}$$

for all the eigenvalues λ *of A.*

Theorem 2 will help us establish the asymptotic convergence of our stabilized system states to zero. The following theorem summarizes our result.

Theorem 3. *The 2D fractional map (6) can be controlled under the 1D control law:*

$$u(t) = -x(t) - a_1 x^2(t) - a_2 y^2(t) + a_3 x(t) y(t) + a_4.\tag{15}$$

Proof. Adding a time-varying control term $u(t)$ to the second state of our fractional map (6) yields:

$$\begin{cases} ^C\Delta_a^v x(t) = y(t - 1 + v) - x(t - 1 + v), \\ ^C\Delta_a^v y(t) = x(t - 1 + v) + a_1 x^2(t - 1 + v) + a_2 y^2(t - 1 + v) \\ \qquad - a_3 x(t - 1 + v) y(t - 1 + v) - a_4 - y(t - 1 + v) \\ \qquad + u(t + v - 1). \end{cases}\tag{16}$$

The aim is to show that with $u(t)$ defined according to (15), the states of (16) converge towards zero asymptotically. In other words, we want to show that point $(0, 0)$ in phase-space is an asymptotically stable equilibrium of the system resulting from substitution of (15) into (16), which is simply:

$$\begin{cases} ^C\Delta_a^v x(t) = -x(t - 1 + v) + y(t - 1 + v), \\ ^C\Delta_a^v y(t) = -y(t - 1 + v). \end{cases}\tag{17}$$

System (17) is linear, which makes our job easy. We write it in matrix form as:

$$^{C}\Delta_a^v \begin{pmatrix} x(t) \\ y(t) \end{pmatrix} = A \begin{pmatrix} x(t) \\ y(t) \end{pmatrix},$$ (18)

where:

$$A = \begin{pmatrix} -1 & 1 \\ 0 & -1 \end{pmatrix}.$$ (19)

Since A is upper-triangular, its eigenvalues are simply $\lambda_1 = \lambda_2 = -1$, and both of them satisfy:

$$|\lambda_i| < \left(2\cos \frac{|\arg \lambda_i| - \pi}{2 - v} \right)^v \text{ and } |\arg \lambda_i| > \frac{v\pi}{2}, \quad i = 1, 2.$$ (20)

Hence, by Theorem 2, we know that the zero equilibrium of (17) is asymptotically stable, and consequently, the states of (16) are stabilized. $\square$

The control strategy proposed in Theorem 3 has been implemented numerically to confirm its validity. The fractional map proposed in this paper was run with initial conditions $(x(0), y(0)) = (0.93, -0.44)$, parameters $(a_1, a_2, a_3, a_4) = (0.2, 0.71, 0.91, 1.14)$ and fractional order $v = 0.97$. The term $u(t)$ defined in (15) was computed iteratively based on previous states and added to the new second state. Figure 9 depicts the time evolution of the states. Clearly, the states converge towards zero, and thus, our stabilization is successful.

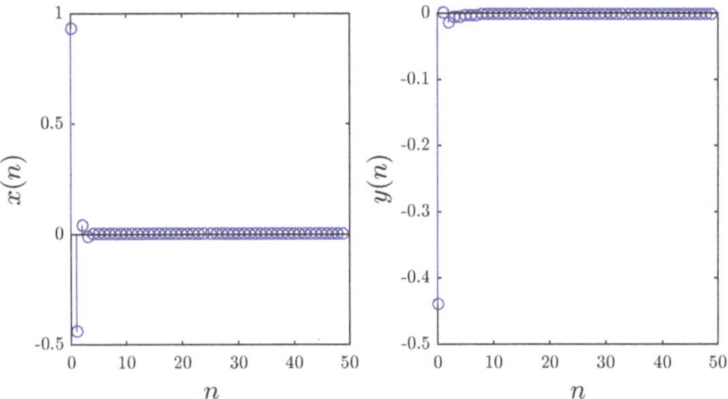

Figure 9. Stabilized states of the fractional map based on control law (15) with $(x(0), y(0)) = (0.93, -0.44)$, $(a_1, a_2, a_3, a_4) = (0.2, 0.71, 0.91, 1.14)$ and $v = 0.97$.

Note that the asymptotic convergence of the control law proposed in Theorem 3 was only established in the commensurate case, i.e., identical fractional orders for all states. However, experimental results have in fact shown that the zero solution of the feedback controlled system is asymptotically stable over a range of circumstances including the incommensurate case and the time varying fractional order case. Figure 10 shows the stabilized states subject to the same parameters and initial setting adopted earlier, but with fractional orders $(v_x, v_y) = (0.95, 0.99)$. The convergence of the states towards zero is apparent. Figure 11 depicts the time evolution of the controlled states when the fractional order is a discrete function of time given by:

$$v(t) = 0.8 + 0.2 \sin\left(\frac{x}{50} \right).$$ (21)

Again, the states converge in a very steady way towards zero, but as expected, take longer to converge compared to the standard case.

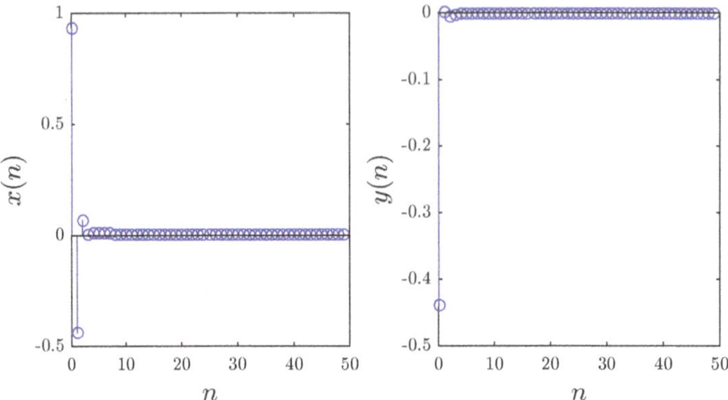

Figure 10. Stabilized states of the fractional map based on control law (15) with $(x(0), y(0)) = (0.93, -0.44)$, $(a_1, a_2, a_3, a_4) = (0.2, 0.71, 0.91, 1.14)$ and $(v_x, v_y) = (0.95, 0.99)$.

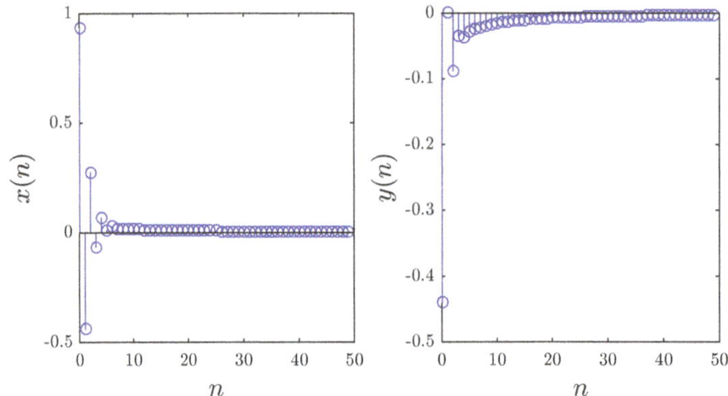

Figure 11. Stabilized states of the fractional map based on control law (15) with $(x(0), y(0)) = (0.93, -0.44)$, $(a_1, a_2, a_3, a_4) = (0.2, 0.71, 0.91, 1.14)$ and time varying fractional order (21).

5. Discussion and Conclusions

In this paper, we have examined a fractional chaotic map based on the standard generic map proposed in [14], which exhibits rich dynamics and hidden chaotic attractors under different circumstances, i.e., with no fixed points, with a single fixed point or with two fixed points. Using phase-space portraits and bifurcation diagrams, we have shown that varying the fractional order impacts the parameter interval over which chaos is observed, as well as the shape of the resulting attractors. We have also quantified the level of chaos present in the proposed map by means of the approximate entropy measure. In addition, we have presented a one-dimensional stabilization controller that forces the system states towards zero asymptotically. Numerical methods were employed to confirm the convergence of this controller under different scenarios. The controller has been shown to be resilient to the time variation of the fractional order. As mentioned before, the standard map upon which our proposed system is based exhibits hidden attractors. This feature has not been assessed for the new fractional map. It is our intention to address this point in a future study.

In addition, it has been claimed in many studies that fractional chaotic maps are superior to their integer counterparts as they involve new degrees of freedom due to the dependence of the maps' dynamics and trajectory on the fractional order [32]. This makes them of particular importance in fields where the seemingly random nature of the dynamical states is a desired property such as in data and image encryption [28]. For instance, we believe that by varying the fractional order by means of some deterministic time series, a new dimension can be introduced into the encryption process, making it harder to break. We have not seen any such studies in the literature. Perhaps the reason for that is the fact that it is rather difficult to establish the convergence of the synchronization errors in such a scenario. However, experiments have shown that many of the proposed controllers in the literature remain valid when the fractional order varies in the same way at the master and slave sides. This will be investigated thoroughly in a future study.

Author Contributions: Conceptualization, A.O. Formal analysis, X.W. and S.B. Investigation, A.-A.K. Methodology, A.O. and V.-T.P. Project administration, F.E.A. Resources, S.B. Software, X.W. and A.-A.K. Writing, original draft, F.E.A. Writing, review and editing, V.-T.P.

Funding: The author Xiong Wang was supported by the National Natural Science Foundation of China (No. 61601306) and Shenzhen Overseas High Level Talent Peacock Project Fund (No. 20150215145C).

Conflicts of Interest: The authors declare no conflict of interest.

References

1. Hénon, M. A two-dimensional mapping with a strange attractor. *Commun. Math. Phys.* **1976**, *50*, 69–77. [CrossRef]
2. Elaydi, S.N. *Discrete Chaos: With Applications in Science and Engineering*, 2nd ed.; Chapman and Hall/CRC: Boca Raton, FL, USA, 2007.
3. Lozi, R. Un atracteur étrange du type attracteur de Hénon. *J. Phys.* **1978**, *39*, 9–10. (In French)
4. Hitzl, D.L.; Zele, F. An exploration of the Hénon quadratic map. *Phys. D Nonlinear Phenom.* **1985**, *14*, 305–326. [CrossRef]
5. Baier, G.G.; Sahle, S. Design of hyperchaotic flows. *Phys. Rev. E* **1995**, *51*, 2712–2714. [CrossRef]
6. Stefanski, K. Modelling chaos and hyperchaos with 3D maps. *Chaos Solitons Fractals* **1998**, *9*, 83–93. [CrossRef]
7. Itoh, M.; Yang, T.; Chua, L.O. Conditions for impulsive synchronization of chaotic and hyperchaotic systems. *Int. J. Bifurc. Chaos Appl. Sci. Eng.* **2001**, *11*, 551–558. [CrossRef]
8. Ouannas, A.; Grassi, G. Inverse full state hybrid projective synchronization for chaotic maps with different dimensions. *Chin. Phys. B* **2016**, *25*, 090503. [CrossRef]
9. Ouannas, A.; Grassi, G.; Karouma, A.; Ziar, T.; Wang, X.; Pham, V.T. New type of chaos synchronization in discrete-time systems: The F-M synchronization. *Open Phys.* **2018**, *16*, 174–182. [CrossRef]
10. Ouannas, A.; Obidat, Z.; Shawagfeh, N. Universal chaos synchronization control laws for general quadratic discrete-time systems. *Appl. Theor. Model* **2017**, *45*, 636–641. [CrossRef]
11. Ouannas, A.; Obidat, Z.; Shawagfeh, N. A new synchronization result for discrete-time chaotic systems. *Differ. Equ. Dyn. Syst.* **2018**, *26*, 125. [CrossRef]
12. Ouannas, A.; Al-Sawalha, M.M. A new approach to synchronize different dimensional chaotic maps using two scaling matrices. *Nonlinear Dyn. Syst. Theory* **2015**, *15*, 400–408.
13. Ouannas, A.; Al-Sawalha, M.M. Synchronization of chaotic dynamical systems in discrete-time. In *Advances in Chaos Theory and Intelligent Control: Studies in Fuziness and Soft Computing*; Azar, A., Vaydiyanathan, S., Eds.; Springer: New York, NY, USA, 2017.
14. Jiang, H.; Liu, Y.; Wei, Z.; Zhang, L. Hidden chaotic attractors in a class of two–dimensional maps. *Nonlinear Dyn.* **2016**, *85*, 2719–2727. [CrossRef]
15. Leonov, G.; Kuznetsov, N.; Vagaitsev, V. Localization of hidden Chua's attractors. *Phys. Lett. A* **2011**, *375*, 2230–2233. [CrossRef]
16. Leonov, G.; Kuznetsov, N.; Vagaitsev, V. Hidden attractor in smooth Chua systems. *Phys. D Nonlinear Phenom.* **2012**, *241*, 1482–1486. [CrossRef]
17. Leonov, G.A.; Kuznetsov, N.V. Hidden attractors in dynamical systems. *Int. J. Bifurc. Chaos* **2013**, *23*, 1330002. [CrossRef]

18. Leonov, G.; Kuznetsov, N.; Kiseleva, M.; Solovyeva, E.; Zaretskiy, A. Hidden oscillations in mathematical model of drilling system actuated by induction motor with a wound rotor. *Nonlinear Dyn.* **2014**, *77*, 277–288. [CrossRef]

19. Danca, M.-F.; Kuznetsov, N. Hidden chaotic sets in a Hopfield neural system. *Chaos Solitons Fractals* **2017**, *103*, 144–150. [CrossRef]

20. Danca, M.-F.; Kuznetsov, N.; Chen, G. Unusual dynamics and hidden attractors of the Rabinovich–Fabrikant system. *Nonlinear Dyn.* **2017**, *88*, 791–805. [CrossRef]

21. Kuznetsov, N.; Leonov, G.; Yuldashev, M.; Yuldashev, R. Hidden attractors in dynamical models of phase-locked loop circuits: Limitations of simulation in MATLAB and SPICE. *Commun. Nonlinear Sci. Numer. Simul.* **2017**, *51*, 39–49.

22. Sharma, P.; Shrimali, M.; Prasad, A.; Kuznetsov, N.; Leonov, G. Control of multistability in hidden attractors. *Eur. Phys. J. Spec. Top.* **2015**, *224*, 1485–1491. [CrossRef]

23. Sharma, P.R.; Shrimali, M.D.; Prasad, A.; Kuznetsov, N.; Leonov, G. Controlling dynamics of hidden attractors. *Int. J. Bifurc. Chaos* **2015**, *25*, 1550061. [CrossRef]

24. Dudkowski, D.; Jafari, S.; Kapitaniak, T.; Kuznetsov, N.V.; Leonov, G.A.; Prasad, A. Hidden attractors in dynamical systems. *Phys. Rep.* **2016**, *637*, 1–50. [CrossRef]

25. Wang, C.; Ding, Q. A new two–Dimensional map with hidden attractors. *Entropy* **2018**, *20*, 322. [CrossRef]

26. Wu, G.C.; Baleanu, D. Discrete fractional logistic map and its chaos. *Nonlinear Dyn.* **2013**, *75*, 283–287. [CrossRef]

27. Hu, T. Discrete Chaos in Fractional Henon Map. *Appl. Math.* **2014**, *5*, 2243–2248. [CrossRef]

28. Wu, G.C.; Baleanu, D. Discrete chaos in fractional delayed logistic maps. *Nonlinear Dyn.* **2015**, *80*, 1697–1703. [CrossRef]

29. Wu, G.; Baleanu, D. Chaos synchronization of the discrete fractional logistic map. *Signal Process.* **2014**, *102*, 96–99. [CrossRef]

30. Wu, G.; Baleanu, D.; Xie, H.; Chen, F. Chaos synchronization of fractional chaotic maps based on the stability condition. *Physica A* **2016**, *460*, 374–383. [CrossRef]

31. Xin, B.; Liu, L.; Hou, G.; Ma, Y. Chaos synchronization of nonlinear fractional discrete dynamical systems via linear control. *Entropy* **2017**, *19*, 351. [CrossRef]

32. Shukla, M.K.; Sharma, B.B. Investigation of chaos in fractional order generalized hyperchaotic Henon map. *Int. J. Electron. Commun.* **2017**, *78*, 265–273. [CrossRef]

33. Goodrich, C.; Peterson, A.C. *Discrete Fractional Calculus*; Springer: New York, NY, USA, 2015.

34. Diaz, J.B.; Olser, T.J. Differences of fractional order. *Math. Comput.* **1974**, *28*, 185–202. [CrossRef]

35. Atici, F.M.; Eloe, P.W. Discrete fractional calculus with the nabla operator. *Electron. J. Qual. Theory Differ. Equ.* **2009**, *3*, 1–12. [CrossRef]

36. Abdeljawad, T. On Riemann and Caputo fractional differences. *Comput. Math. Appl.* **2011**, *62*, 1602–1611. [CrossRef]

37. Cermak, J.; Gyori, I.; Nechvatal, L. On explicit stability conditions for a linear fractional difference system. *Fract. Calc. Appl. Anal.* **2015**, *18*, 651–672. [CrossRef]

38. Baleanu, D.; Wu, G.; Bai, Y.; Chen, F. Stability analysis of Caputo–like discrete fractional systems. *Commun. Nonlinear Sci. Numer. Simul.* **2017**, *48*, 520–530. [CrossRef]

39. Baleanu, D.; Jajarmi, A.; Asad, J.; Blaszczyk, T. The motion of a bead sliding on a wire in fractional sense. *Acta Phys. Pol. A* **2017**, *131*, 1561–1564. [CrossRef]

40. Jajarmi, A.; Hajipour, M.; Mohammadzadeh, E.; Baleanu, D. A new approach for the nonlinear fractional optimal control problems with external persistent disturbances. *J. Frankl. Inst.* **2018**, *335*, 3938–3967. [CrossRef]

41. Anastassiou, G.A. Principles of delta fractional calculus on time scales and inequalities. *Math. Comput. Model.* **2010**, *52*, 556–566. [CrossRef]

42. Eckmann, J.P.; Ruelle, D. Ergodic theory of chaos and strange attractors. *Rev. Mod. Phys.* **2018**, *57*, 617. [CrossRef]

43. Pincus, S.M. Approximate entropy as a measure of system complexity. *Proc. Natl. Acad. Sci. USA* **1991**, *88*, 2297–2301. [CrossRef] [PubMed]

44. Pincus, S. Approximate entropy (ApEn) as a complexity measure. *Chaos Interdiscip. J. Nonlinear Sci.* **1995**, *5*, 110–117. [CrossRef] [PubMed]
45. Xu, G.H.; Shekofteh, Y.; Akgul, A.; Li, C.B.; Panahi, S. A new chaotic system with a self-excited attractor: Entropy measurement, signal encryption, and parameter estimation. *Entropy* **2018**, *20*, 86. [CrossRef]

Article

Tsallis Entropy of Product MV-Algebra Dynamical Systems

Dagmar Markechová [1],* and Beloslav Riečan [2,3]

[1] Department of Mathematics, Faculty of Natural Sciences, Constantine the Philosopher University in Nitra, A. Hlinku 1, SK-949 01 Nitra, Slovakia
[2] Department of Mathematics, Faculty of Natural Sciences, Matej Bel University, Tajovského 40, SK-974 01 Banská Bystrica, Slovakia; beloslav.riecan@umb.sk
[3] Mathematical Institute, Slovak Academy of Sciences, Štefánikova 49, SK-814 73 Bratislava, Slovakia
* Correspondence: dmarkechova@ukf.sk; Tel.: +421-376-408-111; Fax: +421-376-408-020

Received: 17 July 2018; Accepted: 6 August 2018; Published: 9 August 2018

Abstract: This paper is concerned with the mathematical modelling of Tsallis entropy in product MV-algebra dynamical systems. We define the Tsallis entropy of order α, where $\alpha \in (0,1) \cup (1,\infty)$, of a partition in a product MV-algebra and its conditional version and we examine their properties. Among other, it is shown that the Tsallis entropy of order α, where $\alpha > 1$, has the property of sub-additivity. This property allows us to define, for $\alpha > 1$, the Tsallis entropy of a product MV-algebra dynamical system. It is proven that the proposed entropy measure is invariant under isomorphism of product MV-algebra dynamical systems.

Keywords: product MV-algebra; partition; Tsallis entropy; conditional Tsallis entropy; dynamical system

---

1. Introduction

The Shannon entropy [1] is the foundational concept of information theory (cf. [2]). We remind readers that if an experiment has k outcomes with probabilities $p_1, p_2, \ldots, p_k$, then its Shannon entropy is defined as the sum $\sum_{i=1}^{k} s(p_i)$, where $s : [0,1] \to [0,\infty)$ is Shannon's entropy function defined by:

$$s(x) = -x \log x, \tag{1}$$

for every $x \in [0,1]$ ($0 \log 0$ is defined to be 0). Many years later, the Shannon entropy was exploited surprisingly in a completely different field, namely, in dynamical systems. Recall that by a dynamical system in the sense of classical probability theory, we understand a system (Ω, Σ, μ, T), where (Ω, Σ, μ) is a probability space, and $T : \Omega \to \Omega$ is a measure μ preserving transformation. The entropy of dynamical systems was introduced by Kolmogorov and Sinai [3,4] as an invariant for distinguishing them. Namely, if two dynamical systems are isomorphic, then they have the same entropy. In this way Kolmogorov and Sinai showed the existence of non-isomorphic Bernoulli shifts.

The successful using the Kolmogorov-Sinai entropy of dynamical systems has led to an intensive study of alternative entropy measures of dynamical systems. We note that in Reference [5], the concept of logical entropy $H_l(T)$ of a dynamical system (Ω, Σ, μ, T) was introduced and studied. It has been shown that by replacing Shannon's entropy function by the function $l : [0,1] \to [0,\infty)$ defined by:

$$l(x) = x - x^2, \tag{2}$$

for every $x \in [0,1]$, we obtain the results analogous to the case of Kolmogorov-Sinai entropy theory. The logical entropy $H_l(T)$ is invariant under isomorphism of dynamical systems; so it can be used as an alternative instrument for distinguishing them. For some other recently published results regarding the logical entropy measure, we refer, for example, to References [6–15].

In fact, all of the above mentioned studies are possible in the Kolmogorov probability theory based on the modern integration theory. This allows us to describe and study some of the problems associated with uncertainty. In Reference [16], Zadeh presented another approach to uncertainty when he introduced the concept of a fuzzy set. Whereas the Kolmogorov probability applications are based on objective measurements, the Zadeh fuzzy set theory is based on subjective improvements. One of the first Zadeh papers on the fuzzy set theory was devoted to probability of fuzzy sets (cf. [17]), and therefore, the entropy of fuzzy dynamical systems has also been studied (cf. [18–21]). We recall that the fuzzy set is a mapping $f : \Omega \to [0,1]$, hence, the fuzzy partition of Ω is a family of fuzzy sets $A = \{f_1, f_2, \ldots, f_k\}$ such that $\sum_{i=1}^{k} f_i = 1$. Again we can meet the Shannon formula: $H(A) = -\sum_{i=1}^{k} p_i \log p_i$, where $p_i = \int_{\Omega} f_i d\mu$ (cf. [21]).

Anyway the most useful tool for describing multivalued processes is an MV-algebra [22], especially after its Mundici's characterization as an interval in a lattice ordered group (cf. [23,24]). At present, this structure is being investigated by many researchers, and it is natural that there are results also regarding the entropy in this structure; we refer, for instance, to References [25,26]. A probability theory was also investigated on MV-algebras; for a review see Reference [27]. Of course, in some probability problems it is needed to introduce a product on an MV-algebra, an operation outside the corresponding group addition. The operation of a product on an MV-algebra was introduced independently by Riečan [28] from the point of view of probability and Montagna [29] from the point of view of mathematical logic. We note that the notion of product MV-algebra generalizes some families of fuzzy sets; an example of product MV-algebra is a full tribe of fuzzy sets (see e.g., [30]).

The appropriate entropy theory of Shannon and Kolmogorov-Sinai type for the product MV-algebras was created in References [31,32]. The logical entropy, the logical divergence and the logical mutual information of partitions in a product MV-algebra were studied in Reference [8]. In the present paper, we extend the study of entropy in product MV-algebras to the case of Tsallis entropy.

The concept of Tsallis entropy was introduced in 1988 by Constantino Tsallis [33] as a base for generalizing the usual statistical mechanics. In its form it is identical with the Havrda-Charvát structural α—entropy [34], introduced in 1967 in the framework of information theory. If $P = \{p_1, p_2, \ldots, p_k\}$ is a probability distribution, then its Tsallis entropy of order α, where $\alpha \in (0,1) \cup (1, \infty)$, is defined as the number:

$$T_\alpha(P) = \frac{1}{\alpha - 1}\left(1 - \sum_{i=1}^{k} p_i^{\alpha}\right). \tag{3}$$

The entropic index α describes the deviation of Tsallis entropy from the standard Shannon one. Evidently, if we define, for $\alpha \in (0,1) \cup (1, \infty)$, the function $l_\alpha : [0,1] \to [0, \infty)$ by:

$$l_\alpha(x) = \frac{1}{\alpha - 1}(x - x^{\alpha}), \tag{4}$$

for every $x \in [0,1]$, then the Formula (3) can be written in the following form:

$$T_\alpha(P) = \sum_{i=1}^{k} l_\alpha(p_i). \tag{5}$$

Putting $\alpha = 2$ in Equation (5), we obtain:

$$T_2(P) = \sum_{i=1}^{k} l_2(p_i) = \sum_{i=1}^{k} (p_i - p_i^2) = 1 - \sum_{i=1}^{k} p_i^2,$$

which is the logical entropy of a probability distribution $P = \{p_1, p_2, \ldots, p_k\}$ defined and studied in Reference [6].

The Tsallis entropy is the most important quantity among Tsallis' statistics, which form the foundation of nonextensive statistical mechanics of complex systems; for more details, see Reference [35]. The Tsallis statistics are used to describe systems exhibiting long-range correlations, memory, or fractal properties; their applications have been found for a wide range of phenomena in diverse disciplines such as physics, geophysics, chemistry, biology, economics, medicine, etc. [36–48]. They are also applicable to large domains in communication systems (cf. [49]).

In this article we continue studying entropy in a product MV-algebra, by defining and studying the Tsallis entropy of a partition in a product MV-algebra and the Tsallis entropy of product MV-algebra dynamical systems. The rest of the paper is structured as follows. In the following section, preliminaries and related works are given. Our main results are discussed in Sections 3–5. In Section 3, we define and study the Tsallis entropy of a partition in a product MV-algebra. In Section 4, we introduce the concept of conditional Tsallis entropy of partitions in a product MV-algebra and examine its properties. It is shown that the proposed definitions of Tsallis entropy are consistent, in the case of the limit of α going to 1, with the Shannon entropy of partitions studied in Reference [31]. Section 5 is devoted to the study of Tsallis entropy of product MV-algebra dynamical systems. It is proven that the suggested entropy measure is invariant under isomorphism of product MV-algebra dynamical systems. The last section contains a brief summary.

2. Preliminaries

We begin with recalling the definitions of the basic notions and some known results used in the paper. For defining the notion of MV-algebra, various (but of course equivalent) axiom systems were used (see e.g., [28,50,51]). In this paper, the definition of MV-algebra given by Riečan in Reference [52] is used, which is based on the Mundici representation theorem [23,24]. In view of the Mundici theorem, any MV-algebra may be considered to be an interval of an abelian lattice ordered group. Recall that by an abelian lattice ordered group [53], we mean a triplet $(L, +, \leq)$, where $(L, +)$ is an abelian group, $(L, \leq)$ is a partially ordered set being a lattice and, for every $x, y, z \in L$, $x \leq y \implies x + z \leq y + z$.

Definition 1 ([52]). *An MV-algebra is an algebraic structure $\mathcal{A} = (A, \oplus, *, 0, u)$ satisfying the following conditions:*

(i)   *there exists an abelian lattice ordered group $(L, +, \leq)$ such that $A = [0, u] = \{x \in L; 0 \leq x \leq u\}$, where 0 is the neutral element of $(L, +)$ and u is a strong unit of L (i.e., $u \in L$ such that $u > 0$ and to each $x \in L$ there exists a natural number n with the property $x \leq nu$);*

(ii) $\oplus$ *and $*$ are binary operations on A satisfying the following identities: $x \oplus y = (x + y) \wedge u$, $x * y = (x + y - u) \vee 0$.*

Definition 2 ([27]). *A state on an MV-algebra $\mathcal{A} = (A, \oplus, *, 0, u)$ is a mapping $s : A \to [0, 1]$ with the following two properties:*

(i) $s(u) = 1$;

(ii)   *if $x, y \in A$ such that $x + y \leq u$, then $s(x + y) = s(x) + s(y)$.*

Definition 3 ([28]). *A product MV-algebra is an algebraic structure $(A, \oplus, *, \cdot, 0, u)$, where $(A, \oplus, *, 0, u)$ is an MV-algebra and $\cdot$ is an associative and abelian binary operation on A with the following properties:*

(i)   *for every $x \in A$, $u \cdot x = x$;*

(ii)   *if $x, y, z \in A$ such that $x + y \leq u$, then $z \cdot x + z \cdot y \leq u$, and $z \cdot (x + y) = z \cdot x + z \cdot y$.*

In the following text, we will briefly write $(A, \cdot)$ instead of $(A, \oplus, *, \cdot, 0, u)$. A relevant probability theory for the product MV-algebras was developed by Riečan in Reference [54]; the entropy theory of Shannon and Kolmogorov-Sinai type for the product MV-algebras was proposed in References [31,32]. The logical entropy theory for the product MV-algebras was proposed in Reference [8]. We present the main idea and some results of these theories that will be used in the following text.

By a partition in a product MV-algebra $(A, \cdot)$, we mean a k-tuple $X = (x_1, x_2, \ldots, x_k)$ of (not necessarily different) members of A satisfying the condition $x_1 + x_2 + \ldots + x_k = u$.

Let $X = (x_1, x_2, \ldots, x_k)$, and $Y = (y_1, y_2, \ldots, y_l)$ be two partitions in $(A, \cdot)$. We say that Y is a refinement of X, and we write $X \prec Y$, if there exists a partition $\{\beta(1), \beta(2), \ldots, \beta(k)\}$ of the set $\{1, 2, \ldots, l\}$ such that $x_i = \sum_{j \in \beta(i)} y_j$, for $i = 1, 2, \ldots, k$. Further, we define the join $X \vee Y$ of X and Y as an r-tuple (where $r = k \cdot l$) consisting of the members $x_{ij} = x_i \cdot y_j$, $i = 1, 2, \ldots, k$, $j = 1, 2, \ldots, l$. Since $\sum_{i=1}^{k} \sum_{j=1}^{l} x_i \cdot y_j = \left(\sum_{i=1}^{k} x_i \right) \cdot \left(\sum_{j=1}^{l} y_j \right) = u \cdot u = u$, the r-tuple $X \vee Y$ is a partition in $(A, \cdot)$. It represents an experiment consisting of the realization of X and Y.

Example 1. *Consider a probability space (Ω, Σ, μ) and define $A = \{I_E; E \in \Sigma\}$, where $I_E : \Omega \to \{0, 1\}$ stands for the indicator function of the set $E \in \Sigma$. The family A is closed under the product of indicator functions and it is a special case of product MV-algebras. The map $s : A \to [0, 1]$ defined, for every I_E of A, by $s(I_E) = \mu(E)$, is a state on the considered product MV-algebra $(A, \cdot)$. Evidently, if $\{E_1, E_2, \ldots, E_k\}$ is a measurable partition of (Ω, Σ, μ), then the k-tuple $\left(I_{E_1}, I_{E_2}, \ldots, I_{E_k} \right)$ is a partition in the product MV-algebra $(A, \cdot)$.*

Example 2. *Consider a probability space (Ω, Σ, μ) and the family A of all Σ-measurable functions $f : \Omega \to [0, 1]$, so called full tribe of fuzzy sets (cf., e.g., [21,30]). The family A is closed under the natural product of fuzzy sets and it is an important case of product MV-algebras. The mapping $s : A \to [0, 1]$ defined, for every $f \in A$, by the formula $s(f) = \int_{\Omega} f d\mu$, is a state on the product MV-algebra $(A, \cdot)$. The concept of a partition in the product MV-algebra $(A, \cdot)$ Tcoincides with the notion of a fuzzy partition (cf. [21]).*

The definition of entropy of Shannon type of a partition in a product MV-algebra was introduced in [31] as follows.

Definition 4. *Let $X = (x_1, x_2, \ldots, x_k)$ be any partition in a product MV-algebra $(A, \cdot)$ and $s : A \to [0, 1]$ be a state. Then the entropy of X with respect to s is defined by:*

$$H_s(X) = -\sum_{i=1}^{k} s(x_i) \cdot \log s(x_i). \tag{6}$$

If $X = (x_1, x_2, \ldots, x_k)$, and $Y = (y_1, y_2, \ldots, y_l)$ are two partitions in $(A, \cdot)$, then the conditional entropy of X given $y_j \in Y$ is defined by:

$$H_s(X/y_j) = -\sum_{i=1}^{k} s(x_i/y_j) \cdot \log s(x_i/y_j),$$

where:

$$s(x_i/y_j) = \begin{cases} \frac{s(x_i \cdot y_j)}{s(y_j)}, & \text{if } s(y_j) > 0; \\ 0, & \text{if } s(y_j) = 0. \end{cases} \tag{7}$$

The conditional entropy of X given Y is defined by:

$$H_s(X/Y) = \sum_{j=1}^{l} s(y_j) \cdot H_s(X/y_j) = -\sum_{i=1}^{k} \sum_{j=1}^{l} s(x_i \cdot y_j) \cdot \log \frac{s(x_i \cdot y_j)}{s(y_j)}. \tag{8}$$

It is used the standard convention that $0 \log \frac{0}{x} = 0$ if $x \geq 0$. The basis of the logarithm may be any positive real number, but as a rule logarithms to the basis 2 are taken; the entropy is then expressed in bits. If the natural logarithms are taken in the definition, then the entropy is expressed in nats. The entropy of partitions in a product MV-algebra possesses properties corresponding to properties of Shannon's entropy of measurable partitions; more details can be found in Reference [31].

The definition of logical entropy of a partition in a product MV-algebra was introduced in Reference [8] as follows.

Definition 5. *Let* $X = (x_1, x_2, \ldots, x_k)$ *be a partition in a product MV-algebra* $(A, \cdot)$, *and* $s : A \to [0, 1]$ *be a state. Then the logical entropy of* X *with respect to* s *is defined by:*

$$H_l^s(X) = \sum_{i=1}^{k} l(s(x_i)), \tag{9}$$

where $l : [0, 1] \to [0, \infty)$ *is the logical entropy function defined by Equation (2). If* $X = (x_1, x_2, \ldots, x_k)$, *and* $Y = (y_1, y_2, \ldots, y_l)$ *are two partitions in* $(A, \cdot)$, *then the conditional logical entropy of* X *given* Y *is defined by:*

$$H_l^s(X/Y) = \sum_{j=1}^{l} s(y_j)^2 - \sum_{i=1}^{k} \sum_{j=1}^{l} s(x_i \cdot y_j)^2. \tag{10}$$

3. The Tsallis Entropy of Partitions in a Product MV-Algebra

We begin this section with the definition of Tsallis entropy of a partition in a product MV-algebra $(A, \cdot)$, and then we will examine its properties. In the following, we will suppose that $s : A \to [0, 1]$ is a state.

Definition 6. *Let* $X = (x_1, x_2, \ldots, x_k)$ *be a partition in a product MV-algebra* $(A, \cdot)$. *Then we define the Tsallis entropy of order* α, *where* $\alpha \in (0, 1) \cup (1, \infty)$, *of the partition* X *with respect to* s *by:*

$$T_\alpha^s(X) = \frac{1}{\alpha - 1} \left(1 - \sum_{i=1}^{k} s(x_i)^\alpha \right). \tag{11}$$

Remark 1. *Let us consider the function* $l_\alpha : [0, 1] \to [0, \infty)$ *defined by Equation (4). Since* $\sum_{i=1}^{k} s(x_i) = 1$, *the formula (11) can be expressed in the following form:*

$$T_\alpha^s(X) = \sum_{i=1}^{k} l_\alpha(s(x_i)). \tag{12}$$

Evidently, if we put $\alpha = 2$, *the logical entropy* $H_l^s(X)$ *is obtained. It is possible to verify that the function* l_α *is, for every* $\alpha \in (0, 1) \cup (1, \infty)$, *a non-negative function. Namely, if* $\alpha \in (0, 1)$, *then we have* $x^\alpha \geq x$, *for every* $x \in [0, 1]$, *hence* $l_\alpha(x) = \frac{1}{\alpha - 1}(x - x^\alpha) \geq 0$, *for every* $x \in [0, 1]$. *On the other hand, for* $\alpha > 1$, *we have* $x^\alpha \leq x$, *for every* $x \in [0, 1]$, *hence* $l_\alpha(x) = \frac{1}{\alpha - 1}(x - x^\alpha) \geq 0$, *for every* $x \in [0, 1]$.

Example 3. *Let* $(A, \cdot)$ *be any product MV-algebra. Let us consider the partition* $E = (u)$ *representing an experiment resulting in a certain event. Then* $E \prec X$, *for every partition* X *in* $(A, \cdot)$, *and* $T_\alpha^s(E) = l_\alpha(s(u)) = l_\alpha(1) = 0$.

Theorem 1. *Let* $X = (x_1, x_2, \ldots, x_k)$ *be any partition in a product MV-algebra* $(A, \cdot)$. *Then:*

$$0 \leq T_\alpha^s(X) \leq \frac{1}{\alpha - 1}\left(1 - k^{1-\alpha} \right).$$

The equality $T_\alpha^s(X) = \frac{1}{\alpha - 1}(1 - k^{1-\alpha})$ *holds if and only if the state* s *is uniform over* X, *i.e., if and only if* $s(x_i) = \frac{1}{k}$, *for* $i = 1, 2, \ldots, k$.

Proof. The inequality $T_\alpha^s(X) \geq 0$ follows from the non-negativity of function l_α, so it is sufficient to prove the second assertion. We will use the Jensen inequality. It is easy to verify that the function l_α is concave, therefore, applying the Jensen inequality, we have:

$$l_\alpha\left(\frac{1}{k}\sum_{i=1}^{k} s(x_i) \right) \geq \sum_{i=1}^{k} \frac{1}{k} l_\alpha(s(x_i))$$

with the equality if and only if $s(x_1) = s(x_2) = \ldots = s(x_k)$. Since $\sum_{i=1}^{k} s(x_i) = 1$, it follows that:

$$T_\alpha^s(X) = \sum_{i=1}^{k} l_\alpha(s(x_i)) \le k \cdot l_\alpha\left(\frac{1}{k}\sum_{i=1}^{k} s(x_i)\right) = k \cdot l_\alpha\left(\frac{1}{k}\right) = \frac{k}{\alpha-1}\left(\frac{1}{k} - \left(\frac{1}{k}\right)^\alpha\right) = \frac{1}{\alpha-1}\left(1 - k^{1-\alpha}\right).$$

The equality holds if and only if $s(x_1) = s(x_2) = \ldots = s(x_k)$, i.e., if and only if $s(x_i) = \frac{1}{k}$, for $i = 1, 2, \ldots, k$. $\quad\square$

The following propositions will be needed for the proofs of our results.

Proposition 1. *Let X, Y, Z be partitions in a product MV-algebra $(A, \cdot)$. Then:*

(i) $X \prec X \vee Y$;

(ii) $X \prec Y$ *implies* $X \vee Z \prec Y \vee Z$.

Proof. For the proof, see Reference [8]. $\quad\square$

Proposition 2. *Let X, Y, V, Z be partitions in a product MV-algebra $(A, \cdot)$ such that $X \prec Y$, and $V \prec Z$. Then $X \vee V \prec Y \vee Z$.*

Proof. Let $X = (x_1, x_2, \ldots, x_k)$, $Y = (y_1, y_2, \ldots, y_l)$, $V = (v_1, v_2, \ldots, v_m)$, $Z = (z_1, z_2, \ldots, z_n)$, $X \prec Y$, $V \prec Z$. Then there exists a partition $\{\beta(1), \beta(2), \ldots, \beta(k)\}$ of the set $\{1, 2, \ldots, l\}$ such that $x_i = \sum_{j\in\beta(i)} y_j$, for $i = 1, 2, \ldots, k$, and there exists a partition $\{\gamma(1), \gamma(2), \ldots, \gamma(m)\}$ of the set $\{1, 2, \ldots, n\}$ such that $v_r = \sum_{k\in\gamma(r)} z_k$, for $r = 1, 2, \ldots, m$. Put $\delta(i, r) = \{(j, k); j \in \beta(i), k \in \gamma(r)\}$, for $i = 1, 2, \ldots, k, r = 1, 2, \ldots, m$. We get:

$$x_i \cdot v_r = \left(\sum_{j\in\beta(i)} y_j\right) \cdot \left(\sum_{k\in\gamma(r)} z_k\right) = \sum_{(j,k)\in\delta(i,r)} y_j \cdot z_k,$$

for $i = 1, 2, \ldots, k, r = 1, 2, \ldots, m$, what means that $X \vee V \prec Y \vee Z$. $\quad\square$

Proposition 3. *Let $X = (x_1, x_2, \ldots, x_k)$ be a partition in a product MV-algebra $(A, \cdot)$, and $s : A \to [0, 1]$ be a state. Then:*

(i) $\sum_{i=1}^{k} s(x_i \cdot y) = s(y)$, *for every* $y \in A$;

(ii) $\sum_{i=1}^{k} s(x_i / y) = 1$, *for every* $y \in A$ *with* $s(y) > 0$.

Proof. For the proof of the claim (i), see [8]. If $y \in A$ with $s(y) > 0$, then using the previous equality, we obtain:

$$\sum_{i=1}^{k} s(x_i / y) = \frac{1}{s(y)} \sum_{i=1}^{k} s(x_i \cdot y) = \frac{s(y)}{s(y)} = 1. \quad\square$$

Theorem 2. *Let X, Y be partitions in a product MV-algebra $(A, \cdot)$ such that $X \prec Y$. Then $T_\alpha^s(X) \le T_\alpha^s(Y)$.*

Proof. Suppose that $X = (x_1, x_2, \ldots, x_k)$, $Y = (y_1, y_2, \ldots, y_l)$, $X \prec Y$. Then there exists a partition $\{\beta(1), \beta(2), \ldots, \beta(k)\}$ of the set $\{1, 2, \ldots, l\}$ such that $x_i = \sum_{j\in\beta(i)} y_j$, for $i = 1, 2, \ldots, k$. Hence $s(x_i) = s\left(\sum_{j\in\beta(i)} y_j\right) = \sum_{j\in\beta(i)} s(y_j)$, for $i = 1, 2, \ldots, k$. Consider the case when $\alpha \in (1, \infty)$. Then:

$$s(x_i)^\alpha = \left(\sum_{j\in\beta(i)} s(y_j)\right)^\alpha \ge \sum_{j\in\beta(i)} s(y_j)^\alpha,$$

for $i = 1, 2, \ldots, k$. Summing both sides of the above inequality over i, we get:

$$\sum_{i=1}^{k} s(x_i)^\alpha \ge \sum_{i=1}^{k} \sum_{j\in\beta(i)} s(y_j)^\alpha = \sum_{j=1}^{l} s(y_j)^\alpha.$$

In this case we have $\frac{1}{\alpha-1} > 0$, hence:

$$T_\alpha^s(X) = \frac{1}{\alpha-1}\left(1 - \sum_{i=1}^{k} s(x_i)^\alpha\right) \le \frac{1}{\alpha-1}\left(1 - \sum_{j=1}^{l} s(y_j)^\alpha\right) = T_\alpha^s(Y).$$

The case of $\alpha \in (0,1)$ can be obtained in the same way. $\square$

As an immediate consequence of the previous theorem and Proposition 1, we obtain the following result.

Corollary 1. *For every partitions* X, Y *in a product MV-algebra* $(A, \cdot)$, *it holds:*

$$T_\alpha^s(X \vee Y) \ge \max[T_\alpha^s(X), T_\alpha^s(Y)].$$

Proposition 4. *Let* $X = (x_1, x_2, \ldots, x_k)$, *and* $Y = (y_1, y_2, \ldots, y_l)$ *be partitions in a product MV-algebra* $(A, \cdot)$. *Then, for* $\alpha > 1$, *it holds:*

$$\sum_{j=1}^{l} s(y_j)^\alpha \sum_{i=1}^{k} l_\alpha\big(s(x_i/y_j)\big) \le T_\alpha^s(X).$$

Proof. Applying the Jensen inequality, we have:

$$\sum_{j=1}^{l} s(y_j) \cdot l_\alpha\big(s(x_i/y_j)\big) \le l_\alpha\left(\sum_{j=1}^{l} s(y_j) \cdot s(x_i/y_j)\right) = l_\alpha\left(\sum_{j=1}^{l} s(x_i \cdot y_j)\right) = l_\alpha(s(x_i)),$$

for $i = 1, 2, \ldots, k$, and consequently:

$$\sum_{j=1}^{l} s(y_j) \cdot \sum_{i=1}^{k} l_\alpha\big(s(x_i/y_j)\big) \le \sum_{i=1}^{k} l_\alpha(s(x_i)) = T_\alpha^s(X). \tag{13}$$

The assumption that $\alpha > 1$ implies the inequality $s(y_j)^\alpha \le s(y_j)$, for $j = 1, 2, \ldots, l$. The function l_α is non-negative, therefore, for $j = 1, 2, \ldots, l$, we get:

$$s(y_j)^\alpha \sum_{i=1}^{k} l_\alpha\big(s(x_i/y_j)\big) \le s(y_j) \sum_{i=1}^{k} l_\alpha\big(s(x_i/y_j)\big),$$

and consequently:

$$\sum_{j=1}^{l} s(y_j)^\alpha \sum_{i=1}^{k} l_\alpha\big(s(x_i/y_j)\big) \le \sum_{j=1}^{l} s(y_j) \sum_{i=1}^{k} l_\alpha\big(s(x_i/y_j)\big).$$

The last inequality combined with (13) yields the claim. $\square$

Theorem 3. *Let* X, Y *be partitions in a product MV-algebra* $(A, \cdot)$. *Then, for* $\alpha > 1$, *it holds:*

$$T_\alpha^s(X \vee Y) \le T_\alpha^s(X) + T_\alpha^s(Y).$$

Proof. Suppose that $X = (x_1, x_2, \ldots, x_k)$, $Y = (y_1, y_2, \ldots, y_l)$. Let us calculate:

$$T_\alpha^s(X \vee Y) = \frac{1}{\alpha-1}\left(1 - \sum_{i=1}^{k}\sum_{j=1}^{l} s(x_i \cdot y_j)^\alpha\right) = \frac{1}{\alpha-1}\left(1 - \sum_{j=1}^{l} s(y_j)^\alpha \sum_{i=1}^{k} s(x_i / y_j)^\alpha\right)$$

$$= \frac{1}{\alpha-1}\left(1 - \sum_{j=1}^{l} s(y_j)^\alpha + \sum_{j=1}^{l} s(y_j)^\alpha - \sum_{j=1}^{l} s(y_j)^\alpha \sum_{i=1}^{k} s(x_i / y_j)^\alpha\right)$$

$$= \frac{1}{\alpha-1}\left(1 - \sum_{j=1}^{l} s(y_j)^\alpha\right) + \sum_{j=1}^{l} s(y_j)^\alpha \frac{1}{\alpha-1}\left(1 - \sum_{i=1}^{k} s(x_i / y_j)^\alpha\right)$$

$$= T_\alpha^s(Y) + \sum_{j=1}^{l} s(y_j)^\alpha \frac{1}{\alpha-1}\left(\sum_{i=1}^{k} s(x_i / y_j) - \sum_{i=1}^{k} s(x_i / y_j)^\alpha\right)$$

$$= T_\alpha^s(Y) + \sum_{j=1}^{l} s(y_j)^\alpha \frac{1}{\alpha-1}\sum_{i=1}^{k}\left(s(x_i / y_j) - s(x_i / y_j)^\alpha\right)$$

$$= T_\alpha^s(Y) + \sum_{j=1}^{l} s(y_j)^\alpha \sum_{i=1}^{k} l_\alpha\left(s(x_i / y_j)\right) \le T_\alpha^s(X) + T_\alpha^s(Y).$$

In the last step we used Proposition 4. $\square$

Example 4. *Let us consider the family A of all Borel measurable functions $f : [0,1] \to [0,1]$, and define in A the operation $\cdot$ as the natural product of fuzzy sets. Then the system $(A, \cdot)$ is a product MV-algebra. In addition, we define a state $s : A \to [0,1]$ by the formula $s(f) = \int_0^1 f(x)dx$, for every $f \in A$, and consider the pairs $X = (f_1, f_2)$, $Y = (g_1, g_2)$, where $f_1(x) = x$, $f_2(x) = 1 - x$, $g_1(x) = x^2$, $g_2(x) = 1 - x^2$, for every $x \in [0,1]$. Evidently, X and Y are partitions in the product MV-algebra $(A, \cdot)$. By elementary calculations we get that they have the state values $\frac{1}{2}, \frac{1}{2}$ and $\frac{1}{3}, \frac{2}{3}$ of the corresponding elements, respectively. The partition $X \vee Y = (f_1 \cdot g_1, f_1 \cdot g_2, f_2 \cdot g_1, f_2 \cdot g_2)$ has the state values $\frac{1}{4}, \frac{1}{4}, \frac{1}{12}, \frac{5}{12}$ of the corresponding elements. We want to find out whether the statement of the previous theorem is true in the case under consideration. Using the formula (11), it can be computed that $T_2^s(X) = 0.5$, $T_2^s(Y) \doteq 0.4444$, $T_2^s(X \vee Y) \doteq 0.6944$, $T_3^s(X) = 0.375$, $T_3^s(Y) \doteq 0.3333$, $T_3^s(X \vee Y) \doteq 0.4479$. It holds $T_2^s(X \vee Y) < T_2^s(X) + T_2^s(Y)$, and $T_3^s(X \vee Y) < T_3^s(X) + T_3^s(Y)$, which is consistent with the assertion of Theorem 3. Put $\alpha = \frac{1}{2}$. We obtain: $T_{1/2}^s(X) \doteq 0.8284$, $T_{1/2}^s(Y) \doteq 0.7877$, $T_{1/2}^s(X \vee Y) \doteq 1.8683$. It can be seen that $T_{1/2}^s(X \vee Y) > T_{1/2}^s(X) + T_{1/2}^s(Y)$. The result means that the Tsallis entropy $T_\alpha^s(X)$ of order $\alpha \in (0,1)$ does not have the property of sub-additivity.*

One of the most important properties of Shannon entropy is additivity. In the following theorem it is shown that the Tsallis entropy $T_\alpha^s(X)$ does not have the property of additivity; it satisfies the following weaker property of pseudo-additivity.

Theorem 4. *If partitions X, Y in a product MV-algebra $(A, \cdot)$ are statistically independent with respect to s, i.e., $s(x \cdot y) = s(x) \cdot s(y)$, for every $x \in X$, and $y \in Y$, then:*

$$T_\alpha^s(X \vee Y) = T_\alpha^s(X) + T_\alpha^s(Y) + (1 - \alpha) \cdot T_\alpha^s(X) \cdot T_\alpha^s(Y).$$

Proof. Suppose that $X = (x_1, x_2, \ldots, x_k)$, $Y = (y_1, y_2, \ldots, y_l)$. Let us calculate:

$$T_\alpha^s(X \vee Y) = \frac{1}{\alpha-1}\left(1 - \sum_{i=1}^{k}\sum_{j=1}^{l} s(x_i \cdot y_j)^\alpha\right) = \frac{1}{\alpha-1}\left(1 - \sum_{i=1}^{k} s(x_i)^\alpha \sum_{j=1}^{l} s(y_j)^\alpha\right)$$

$$= \frac{1}{\alpha-1}\left(1 - \sum_{j=1}^{l} s(y_j)^\alpha + \sum_{j=1}^{l} s(y_j)^\alpha - \sum_{i=1}^{k} s(x_i)^\alpha \sum_{j=1}^{l} s(y_j)^\alpha\right)$$

$$= \frac{1}{\alpha-1}\left(1 - \sum_{j=1}^{l} s(y_j)^\alpha\right) + \frac{1}{\alpha-1}\sum_{j=1}^{l} s(y_j)^\alpha \left(1 - \sum_{i=1}^{k} s(x_i)^\alpha\right)$$

$$= T_\alpha^s(Y) + T_\alpha^s(X) + (1 - \alpha) \cdot T_\alpha^s(X) \cdot T_\alpha^s(Y). \square$$

In the last part of this section, we will prove the concavity of Tsallis entropy $T_\alpha^s(X)$ on the family of all states defined on a given product MV-algebra $(A, \cdot)$.

Proposition 5. *Let s_1, s_2 be two states defined on a common product MV-algebra $(A, \cdot)$. Then, for every real number $\lambda \in [0, 1]$, the map $\lambda s_1 + (1 - \lambda)s_2 : A \to [0, 1]$ is a state on $(A, \cdot)$.*

Proof. The proof is simple, so it is omitted. $\square$

Theorem 5. *Let s_1, s_2 be two states defined on a common product MV-algebra $(A, \cdot)$. Then, for every partition X in a product MV-algebra $(A, \cdot)$, and for every real number $\lambda \in [0, 1]$, the following inequality holds:*

$$\lambda T_\alpha^{s_1}(X) + (1 - \lambda)T_\alpha^{s_2}(X) \leq T_\alpha^{\lambda s_1 + (1 - \lambda)s_2}(X).$$

Proof. Assume that $X = (x_1, x_2, \ldots, x_k)$. The function l_α is concave, therefore, for every real number $\lambda \in [0, 1]$, we get:

$$\lambda T_\alpha^{s_1}(X) + (1 - \lambda)T_\alpha^{s_2}(X) = \lambda \sum_{i=1}^{k} l_\alpha(s_1(x_i)) + (1 - \lambda) \sum_{i=1}^{k} l_\alpha(s_2(x_i))$$

$$= \sum_{i=1}^{k} (\lambda l_\alpha(s_1(x_i)) + (1 - \lambda)l_\alpha(s_2(x_i))) \leq \sum_{i=1}^{k} l_\alpha(\lambda s_1(x_i) + (1 - \lambda)s_2(x_i))$$

$$= \sum_{i=1}^{k} l_\alpha((\lambda s_1 + (1 - \lambda)s_2)(x_i)) = T_\alpha^{\lambda s_1 + (1 - \lambda)s_2}(X).$$

As a consequence of Theorem 5, we get the concavity of the logical entropy $H_l^s(X)$ as a function of s. The result of the previous theorem is illustrated in the following example.

Example 5. *Consider the product MV-algebra $(A, \cdot)$ from Example 4 and the real functions G_1, G_2 defined by the equalities $G_1(x) = x, G_2(x) = x^2$, for every real number x. We define two states $s_1 : A \to [0, 1]$, $s_2 : A \to [0, 1]$ by the formulas $s_1(f) = \int_0^1 f(x)dG_1(x) = \int_0^1 f(x)dx$, $s_2(f) = \int_0^1 f(x)dG_2(x) = \int_0^1 f(x)2x\,dx$, for every f of A. Further, we consider the partition $X = \left(I_{[0,\frac{1}{3})}, I_{[\frac{1}{3},1]}\right)$ in $(A, \cdot)$. By simple calculation we get that it has the s_1-state values $\frac{1}{3}, \frac{2}{3}$ of the corresponding elements, and the s_2-state values $\frac{1}{9}, \frac{8}{9}$ of the corresponding elements. In the previous theorem we put $\lambda = 0.2$. We will show that, for the chosen $\alpha \in (0, 1) \cup (1, \infty)$, the following inequality holds:*

$$0.2 \cdot T_\alpha^{s_1}(X) + 0.8 \cdot T_\alpha^{s_2}(X) \leq T_\alpha^{0.2s_1 + 0.8s_2}(X). \tag{14}$$

Put $\alpha = \frac{1}{2}$. We calculated that $T_{1/2}^{s_1}(X) \doteq 0.7877$, $T_{1/2}^{s_2}(X) \doteq 0.5523$, and $T_{1/2}^{0.2s_1 + 0.8s_2}(X) \doteq 0.6267$. One can easily check that in this case:

$$0.2 \cdot T_{1/2}^{s_1}(X) + 0.8 \cdot T_{1/2}^{s_2}(X) < T_{1/2}^{0.2s_1 + 0.8s_2}(X).$$

For the case of $\alpha = 2$, i.e., for the logical entropy, we get: $T_2^{s_1}(X) \doteq 0.4444$, $T_2^{s_2}(X) \doteq 0.1975$, $T_2^{0.2s_1 + 0.8s_2}(X) \doteq 0.2627$, and for the case of $\alpha = 3$, we obtain: $T_3^{s_1}(X) \doteq 0.3333$, $T_3^{s_2}(X) \doteq 0.148148$, $T_3^{0.2s_1 + 0.8s_2}(X) \doteq 0.19704$. One can easily check that in both cases the inequality (14) holds.

4. The Conditional Tsallis Entropy of Partitions in a Product MV-Algebra

In this section we introduce and study the concept of conditional Tsallis entropy of partitions in a product MV-algebra $(A, \cdot)$.

Definition 7. *Let $X = (x_1, x_2, \ldots, x_k)$, and $Y = (y_1, y_2, \ldots, y_l)$ be partitions in a product MV-algebra $(A, \cdot)$. We define the conditional Tsallis entropy of order α, where $\alpha \in (0, 1) \cup (1, \infty)$, of X given Y as the number:*

$$T_\alpha^s(X/Y) = \frac{1}{\alpha - 1}\left(\sum_{j=1}^l s(y_j)^\alpha - \sum_{i=1}^k \sum_{j=1}^l s(x_i \cdot y_j)^\alpha\right). \tag{15}$$

Remark 2. *Evidently, if we put $\alpha = 2$, then we obtain the conditional logical entropy of X given Y defined by Equation (10).*

At $\alpha = 1$ the value of $T_\alpha^s(X/Y)$ is undefined because it gives the shape $\frac{0}{0}$. In the following theorem it is shown that for $\alpha \to 1$ the conditional Tsallis entropy $T_\alpha^s(X/Y)$ tends to the conditional Shannon entropy $H_s(X/Y)$ defined by the formula (8), when the natural logarithm is taken in this formula.

Theorem 6. *Let $X = (x_1, x_2, \ldots, x_k)$, and $Y = (y_1, y_2, \ldots, y_l)$ be partitions in a product MV-algebra $(A, \cdot)$. Then:*

$$\lim_{\alpha \to 1} T_\alpha^s(X/Y) = -\sum_{i=1}^k \sum_{j=1}^l s(x_i \cdot y_j) \cdot \ln \frac{s(x_i \cdot y_j)}{s(y_j)}.$$

Proof. For every $\alpha \in (0, 1) \cup (1, \infty)$, we have:

$$T_\alpha^s(X/Y) = \frac{1}{\alpha - 1}\left(\sum_{j=1}^l s(y_j)^\alpha - \sum_{i=1}^k \sum_{j=1}^l s(x_i \cdot y_j)^\alpha\right) = \frac{f(\alpha)}{g(\alpha)},$$

where f and g are continuous functions defined, for every $\alpha \in (0, \infty)$, by the equalities:

$$f(\alpha) = \sum_{j=1}^l s(y_j)^\alpha - \sum_{i=1}^k \sum_{j=1}^l s(x_i \cdot y_j)^\alpha, g(\alpha) = \alpha - 1.$$

The functions f and g are differentiable and evidently, $\lim_{\alpha \to 1} g(\alpha) = 0$. Also, it can easily be verified that $\lim_{\alpha \to 1} f(\alpha) = 0$. Indeed, by Proposition 3, we get:

$$\lim_{\alpha \to 1} f(\alpha) = \sum_{j=1}^l s(y_j) - \sum_{i=1}^k \sum_{j=1}^l s(x_i \cdot y_j) = 1 - \sum_{i=1}^k s(x_i) = 1 - 1 = 0.$$

Using L'Hôpital's rule, it follows that $\lim_{\alpha \to 1} T_\alpha^s(X/Y) = \lim_{\alpha \to 1} \frac{f'(\alpha)}{g'(\alpha)}$, under the assumption that the right hand side exists. It holds $\frac{d}{d\alpha} g(\alpha) = 1$, and:

$$\frac{d}{d\alpha} f(\alpha) = \sum_{j=1}^l \frac{d}{d\alpha}\left(s(y_j)^\alpha\right) - \sum_{i=1}^k \sum_{j=1}^l \frac{d}{d\alpha}\left(s(x_i \cdot y_j)^\alpha\right) = \sum_{j=1}^l s(y_j)^\alpha \ln s(y_j) - \sum_{i=1}^k \sum_{j=1}^l s(x_i \cdot y_j)^\alpha \ln s(x_i \cdot y_j).$$

It follows that:

$$\lim_{\alpha \to 1} T_\alpha^s(X/Y) = \lim_{\alpha \to 1} f'(\alpha) = \sum_{j=1}^l s(y_j) \ln s(y_j) - \sum_{i=1}^k \sum_{j=1}^l s(x_i \cdot y_j) \ln s(x_i \cdot y_j)$$

$$= \sum_{i=1}^k \sum_{j=1}^l s(x_i \cdot y_j) \cdot \ln s(y_j) - \sum_{i=1}^k \sum_{j=1}^l s(x_i \cdot y_j) \cdot \ln s(x_i \cdot y_j) = -\sum_{i=1}^k \sum_{j=1}^l s(x_i \cdot y_j) \cdot \ln \frac{s(x_i \cdot y_j)}{s(y_j)}. \quad \square$$

Example 6. *Let $X = (x_1, x_2, \ldots, x_k)$ be any partition in a product MV-algebra $(A, \cdot)$, and $E = (u)$. Then:*

$$T_\alpha^s(X/E) = \frac{1}{\alpha - 1}\left(s(u)^\alpha - \sum_{i=1}^k s(x_i \cdot u)^\alpha\right) = \frac{1}{\alpha - 1}\left(1 - \sum_{i=1}^k s(x_i)^\alpha\right) = T_\alpha^s(X).$$

Theorem 7. *Let* $X = (x_1, x_2, \ldots, x_k)$ *be any partition in a product MV-algebra* $(A, \cdot)$. *Then:*

$$\lim_{\alpha \to 1} T_\alpha^s(X) = -\sum_{i=1}^{k} s(x_i) \cdot \ln s(x_i).$$

Proof. The statement is an immediate consequence of the previous theorem; it suffices to put $Y = E = (u)$. $\square$

Theorem 8. *For arbitrary partitions* X, Y, Z *in a product MV-algebra* $(A, \cdot)$, *it holds:*

(i) $T_\alpha^s(X/Y) \geq 0;$
(ii) $T_\alpha^s(X \vee Y/Z) = T_\alpha^s(X/Z) + T_\alpha^s(Y/X \vee Z);$
(iii) $T_\alpha^s(X \vee Y) = T_\alpha^s(X) + T_\alpha^s(Y/X).$

Proof. Let $X = (x_1, x_2, \ldots, x_k)$, $Y = (y_1, y_2, \ldots, y_l)$, $Z = (z_1, z_2, \ldots, z_m)$.

(i)   By Proposition 3, it holds $s(y_j) = \sum_{i=1}^{k} s(x_i \cdot y_j)$, for $j = 1, 2, \ldots, l$, hence, we can write:

$$T_\alpha^s(X/Y) = \tfrac{1}{\alpha-1}\left(\sum_{j=1}^{l} s(y_j)^\alpha - \sum_{i=1}^{k}\sum_{j=1}^{l} s(x_i \cdot y_j)^\alpha\right) = \tfrac{1}{\alpha-1}\left(\sum_{j=1}^{l} s(y_j)^{\alpha-1}\sum_{i=1}^{k} s(x_i \cdot y_j) - \sum_{i=1}^{k}\sum_{j=1}^{l} s(x_i \cdot y_j)^\alpha\right)$$

$$= \tfrac{1}{\alpha-1}\sum_{i=1}^{k}\sum_{j=1}^{l} s(x_i \cdot y_j)\left(s(y_j)^{\alpha-1} - s(x_i \cdot y_j)^{\alpha-1}\right).$$

Suppose that $\alpha \in (1, \infty)$. For $i = 1, 2, \ldots, k$, $j = 1, 2, \ldots, l$, we have $s(x_i \cdot y_j) \leq s(y_j)$, which implies that $s(x_i \cdot y_j)^{\alpha-1} \leq s(y_j)^{\alpha-1}$, for $i = 1, 2, \ldots, k$, $j = 1, 2, \ldots, l$. Since $\tfrac{1}{\alpha-1} > 0$, for $\alpha \in (1, \infty)$, it follows that $T_\alpha^s(X/Y) \geq 0$. On the other hand, for $\alpha \in (0, 1)$, it holds $s(x_i \cdot y_j)^{\alpha-1} \geq s(y_j)^{\alpha-1}$, for $i = 1, 2, \ldots, k$, $j = 1, 2, \ldots, l$. In this case $\tfrac{1}{\alpha-1} < 0$, hence $T_\alpha^s(X/Y) \geq 0$.

(ii) By direct calculations, we have:

$$T_\alpha^s(X/Z) + T_\alpha^s(Y/X \vee Z) = \tfrac{1}{\alpha-1}\left(\sum_{k=1}^{m} s(z_k)^\alpha - \sum_{i=1}^{k}\sum_{k=1}^{m} s(x_i \cdot z_k)^\alpha\right)$$

$$+ \tfrac{1}{\alpha-1}\left(\sum_{i=1}^{k}\sum_{k=1}^{m} s(x_i \cdot z_k)^\alpha - \sum_{i=1}^{k}\sum_{j=1}^{l}\sum_{k=1}^{m} s(x_i \cdot y_j \cdot z_k)^\alpha\right)$$

$$= \tfrac{1}{\alpha-1}\left(\sum_{k=1}^{m} s(z_k)^\alpha - \sum_{i=1}^{k}\sum_{j=1}^{l}\sum_{k=1}^{m} s(x_i \cdot y_j \cdot z_k)^\alpha\right) = T_\alpha^s(X \vee Y/Z).$$

(iii) The statement is an immediate consequence of the previous property; it suffices to put $Z = E = (u)$. $\square$

By combining the property (iii) from Theorem 8 with Theorem 4, we obtain the following property of conditional Tsallis entropy $T_\alpha^s(X/Y)$.

Theorem 9. *If partitions* X, Y *in a product MV-algebra* $(A, \cdot)$ *are statistically independent with respect to* s, *then:*

$$T_\alpha^s(X/Y) = T_\alpha^s(X) + (1 - \alpha) \cdot T_\alpha^s(X) \cdot T_\alpha^s(Y).$$

Theorem 10. *Let* X, Y *be partitions in a product MV-algebra* $(A, \cdot)$. *Then, for* $\alpha > 1$, *it holds:*

$$T_\alpha^s(X/Y) \leq T_\alpha^s(X).$$

Proof. Let $\alpha > 1$. Then by the use of the property (iii) from Theorem 8 and Theorem 3, we get:

$$T_\alpha^s(X/Y) = T_\alpha^s(X \vee Y) - T_\alpha^s(Y) \leq T_\alpha^s(X) + T_\alpha^s(Y) - T_\alpha^s(Y) = T_\alpha^s(X). \quad \square$$

To illustrate the result of previous theorem, we provide the following example, which is a continuation of Example 4.

Example 7. *Consider the product MV-algebra* $(A, \cdot)$, *the state* $s : A \to [0,1]$ *and the partitions* X, Y *from Example 4. We have calculated that* $T_3^s(X) = 0.375$, $T_3^s(Y) \doteq 0.3333$, $T_{1/2}^s(X) \doteq 0.8284$, $T_{1/2}^s(Y) \doteq 0.7877$. *By easy calculations we get that* $T_3^s(X/Y) \doteq 0.1146$, $T_{1/2}^s(X/Y) \doteq 1.0806$, $T_3^s(Y/X) = 0.0729$, $T_{1/2}^s(Y/X) \doteq 1.0399$. *It can be seen that* $T_3^s(X/Y) < T_3^s(X)$, *and* $T_3^s(Y/X) < T_3^s(Y)$, *which is consistent with the assertion of Theorem 10. On the other hand, we have* $T_{1/2}^s(X/Y) > T_{1/2}^s(X)$, *and* $T_{1/2}^s(Y/X) > T_{1/2}^s(Y)$. *The result means that the conditional Tsallis entropy* $T_\alpha^s(X/Y)$ *of order* $\alpha \in (0,1)$ *does not have the property of monotonicity.*

5. The Tsallis Entropy of Dynamical Systems in a Product MV-Algebra

In this section, we introduce and study the concept of the Tsallis entropy of a dynamical system in a product MV-algebra $(A, \cdot)$.

Definition 8. ([32]). *By a dynamical system in a product MV-algebra* $(A, \cdot)$, *we understand a system* (A, s, τ), *where* $s : A \to [0,1]$ *is a state, and* $\tau : A \to A$ *is a map such that* $\tau(u) = u$, *and, for every* $x, y \in A$, *the following conditions are satisfied:*

(i)   if $x + y \leq u$, *then* $\tau(x) + \tau(y) \leq u$, *and* $\tau(x + y) = \tau(x) + \tau(y)$;
(ii) $\tau(x \cdot y) = \tau(x) \cdot \tau(y)$;
(iii) $s(\tau(x)) = s(x)$.

Remark 3. *We say also briefly a product MV-algebra dynamical system instead of a dynamical system in a product MV-algebra.*

Example 8. *Let* (Ω, Σ, μ, T) *be a classical dynamical system. Let us consider the product MV-algebra* $(A, \cdot)$ *and the state* $s : A \to [0,1]$ *from Example 1. In addition, let us define the mapping* $\tau : A \to A$ *by the equality* $\tau(I_E) = I_E \circ T = I_{T^{-1}(E)}$, *for every* $I_E \in A$. *Then the system* (A, s, τ) *is a dynamical system in the considered product MV-algebra* $(A, \cdot)$.

Example 9. *Let* (Ω, Σ, μ, T) *be a classical dynamical system. Let us consider the product MV-algebra* $(A, \cdot)$ *and the state* $s : A \to [0,1]$ *from Example 2. If we define the mapping* $\tau : A \to A$ *by the equality* $\tau(f) = f \circ T$, *for every* $f \in A$, *then it is easy to verify that the system* (A, s, τ) *is a dynamical system in the considered product MV-algebra* $(A, \cdot)$.

Let (A, s, τ) be a dynamical system in a product MV-algebra $(A, \cdot)$, and $X = (x_1, x_2, \ldots, x_k)$ be a partition in $(A, \cdot)$. Put $\tau(X) = (\tau(x_1), \tau(x_2), \ldots, \tau(x_k))$. Since $x_1 + x_2 + \ldots + x_k = u$, according to Definition 8, we have $\tau(x_1) + \tau(x_2) + \ldots + \tau(x_k) = \tau(x_1 + x_2 + \ldots + x_k) = \tau(u) = u$, what means that the k-tuple $\tau(X)$ is a partition in $(A, \cdot)$.

Proposition 6. *Let* (A, s, τ) *be a dynamical system in a product MV-algebra* $(A, \cdot)$, *and* X, Y *be partitions in* $(A, \cdot)$. *Then*

(i) $\tau(X \vee Y) = \tau(X) \vee \tau(Y)$;
(ii) $X \prec Y$ implies $\tau(X) \prec \tau(Y)$.

Proof. The property (i) follows from the condition (ii) of Definition 8. Suppose that $X = (x_1, x_2, \ldots, x_k)$, $Y = (y_1, y_2, \ldots, y_l)$, $X \prec Y$. Then there exists a partition $\{\beta(1), \beta(2), \ldots, \beta(k)\}$ of the set $\{1, 2, \ldots, l\}$ such that $x_i = \sum_{j \in \beta(i)} y_j$, for Therefore, by the condition (i) from Definition 8, we have:

$$\tau(x_i) = \tau\left(\sum_{j \in \beta(i)} y_j\right) = \sum_{j \in \beta(i)} \tau(y_j), \text{ for } i = 1, 2, \ldots, k.$$

However, this means that $\tau(X) \prec \tau(Y)$. $\quad\square$

Define $\tau^2 = \tau \circ \tau$, and put $\tau^n = \tau \circ \tau^{n-1}$, for $n = 1, 2, \ldots$, where τ^0 is the identical mapping. It is obvious that the mapping $\tau^n : A \to A$ possesses the properties from Definition 8. Hence, for any non-negative integer n, the system (A, s, τ^n) is a dynamical system in a product MV-algebra $(A, \cdot)$.

Theorem 11. *Let (A, s, τ) be a dynamical system in a product MV-algebra $(A, \cdot)$, and X, Y be partitions in $(A, \cdot)$. Then, for any non-negative integer n, the following equalities hold:*

(i) $T_\alpha^s(\tau^n(X)) = T_\alpha^s(X)$;
(ii) $T_\alpha^s(\tau^n(X)/\tau^n(Y)) = T_\alpha^s(X/Y)$.

Proof. Suppose that $X = (x_1, x_2, \ldots, x_k)$, $Y = (y_1, y_2, \ldots, y_l)$.

(i)   Since for any non-negative integer n, and $i = 1, 2, \ldots, k$, it holds $s(\tau^n(x_i)) = s(x_i)$, we obtain:

$$T_\alpha^s(\tau^n(X)) = \sum_{i=1}^{k} l_\alpha\left(s(\tau^n(x_i))\right) = \sum_{i=1}^{k} l_\alpha\left(s(x_i)\right) = T_\alpha^s(X).$$

(ii)   Based on the same argument, we get:

$$T_\alpha^s(\tau^n(X)/\tau^n(Y)) = \tfrac{1}{\alpha-1}\left(\sum_{j=1}^{l} s(\tau^n(y_j))^\alpha - \sum_{i=1}^{k}\sum_{j=1}^{l} s(\tau^n(x_i \cdot y_j))^\alpha\right)$$
$$= \tfrac{1}{\alpha-1}\left(\sum_{j=1}^{l} s(y_j)^\alpha - \sum_{i=1}^{k}\sum_{j=1}^{l} s(x_i \cdot y_j)^\alpha\right) = T_\alpha^s(X/Y).$$

Theorem 12. *Let (A, s, τ) be a dynamical system in a product MV-algebra $(A, \cdot)$, and X be a partition in $(A, \cdot)$. Then, for $n = 2, 3, \ldots$, the following equality holds:*

$$T_\alpha^s\left(\vee_{k=0}^{n-1} \tau^k(X)\right) = T_\alpha^s(X) + \sum_{i=1}^{n-1} T_\alpha^s\left(X/\vee_{k=1}^{i} \tau^k(X)\right).$$

Proof. We use proof by mathematical induction on n, starting with $n = 2$. For $n = 2$, the claim holds by the property (iii) of Theorem 8. We suppose that the claim holds for a given integer $n > 1$, and we will prove that it holds for $n + 1$. By the property (i) of Theorem 11, we get:

$$T_\alpha^s\left(\vee_{k=1}^{n} \tau^k(X)\right) = T_\alpha^s\left(\tau\left(\vee_{k=0}^{n-1} \tau^k(X)\right)\right) = T_\alpha^s\left(\vee_{k=0}^{n-1} \tau^k(X)\right).$$

Therefore, using the property (iii) of Theorem 8 and our inductive hypothesis, we obtain:

$$T_\alpha^s\left(\vee_{k=0}^{n} \tau^k(X)\right) = T_\alpha^s\left(\left(\vee_{k=1}^{n} \tau^k(X)\right) \vee X\right) = T_\alpha^s\left(\vee_{k=1}^{n} \tau^k(X)\right) + T_\alpha^s\left(X/\vee_{k=1}^{n} \tau^k(X)\right)$$
$$= T_\alpha^s\left(\vee_{k=0}^{n-1} \tau^k(X)\right) + T_\alpha^s\left(X/\vee_{k=1}^{n} \tau^k(X)\right)$$
$$= T_\alpha^s(X) + \sum_{i=1}^{n-1} T_\alpha^s\left(X/\vee_{k=1}^{i} \tau^k(X)\right) + T_\alpha^s\left(X/\vee_{k=1}^{n} \tau^k(X)\right)$$
$$= T_\alpha^s(X) + \sum_{i=1}^{n} T_\alpha^s\left(X/\vee_{k=1}^{i} \tau^k(X)\right).$$

In conclusion, the claim is obtained by the principle of mathematical induction. $\quad\square$

In the following, we will define the Tsallis entropy of a dynamical system (A, s, τ). First, we define the Tsallis entropy of τ relative to a partition X in $(A, \cdot)$. Then we remove the dependence on X to get the Tsallis entropy of a dynamical system (A, s, τ). The following proposition will be needed.

Proposition 7. *Let (A, s, τ) be a dynamical system in a product MV-algebra $(A, \cdot)$, and X be a partition in $(A, \cdot)$. Then, for $\alpha > 1$, there exists the following limit:*

$$\lim_{n \to \infty} \frac{1}{n} T_\alpha^s(\vee_{k=0}^{n-1} \tau^k(X)).$$

Proof. Put $c_n = T_\alpha^s(\vee_{k=0}^{n-1} \tau^k(X))$, for $n = 1, 2, \ldots$. Then the sequence $\{c_n\}_{n=1}^\infty$ is a sequence of non-negative real numbers with the property $c_{r+s} \leq c_r + c_s$, for every natural numbers r, s. Indeed, by means of sub-additivity of Tsallis entropy $T_\alpha^s(X)$ for $\alpha > 1$, and the property (i) from Theorem 11, we have:

$$c_{r+s} = T_\alpha^s(\vee_{k=0}^{r+s-1} \tau^k(X)) \leq T_\alpha^s(\vee_{k=0}^{r-1} \tau^k(X)) + T_\alpha^s(\vee_{k=r}^{r+s-1} \tau^k(X))$$
$$= c_r + T_\alpha^s(\tau^r(\vee_{k=0}^{s-1} \tau^k(X))) = c_r + T_\alpha^s(\vee_{k=0}^{s-1} \tau^k(X)) = c_r + c_s.$$

The result guarantees (in view of Theorem 4.9, [55]) the existence of $\lim_{n \to \infty} \frac{1}{n} c_n$. $\square$

Definition 9. *Let (A, s, τ) be a dynamical system in a product MV-algebra $(A, \cdot)$, and X be a partition in $(A, \cdot)$. Then we define, for $\alpha > 1$, the Tsallis entropy of τ relative to X by:*

$$T_\alpha^s(\tau, X) = \lim_{n \to \infty} \frac{1}{n} T_\alpha^s(\vee_{k=0}^{n-1} \tau^k(X)).$$

Remark 4. *Consider any dynamical system (A, s, τ) in a product MV-algebra $(A, \cdot)$, and the partition $E = (u)$. Evidently, $\vee_{k=0}^{n-1} \tau^k(E) = E$, and $T_\alpha^s(\tau, E) = \lim_{n \to \infty} \frac{1}{n} T_\alpha^s(\vee_{k=0}^{n-1} \tau^k(E)) = \lim_{n \to \infty} \frac{1}{n} T_\alpha^s(E) = 0$.*

Theorem 13. *Let (A, s, τ) be a dynamical system in a product MV-algebra $(A, \cdot)$, and X be a partition in $(A, \cdot)$. Then, for $\alpha > 1$, and for any non-negative integer r, the following equality holds:*

$$T_\alpha^s(\tau, X) = T_\alpha^s(\tau, \vee_{i=0}^r \tau^i(X)).$$

Proof. Using Definition 9, we can write:

$$T_\alpha^s(\tau, \vee_{i=0}^r \tau^i(X)) = \lim_{n \to \infty} \frac{1}{n} T_\alpha^s(\vee_{k=0}^{n-1} \tau^k(\vee_{i=0}^r \tau^i(X)))$$
$$= \lim_{n \to \infty} \frac{r+n}{n} \cdot \frac{1}{r+n} T_\alpha^s(\vee_{k=0}^{r+n-1} \tau^k(X))$$
$$= \lim_{n \to \infty} \frac{1}{r+n} T_\alpha^s(\vee_{k=0}^{r+n-1} \tau^k(X)) = T_\alpha^s(\tau, X). \ \square$$

Theorem 14. *Let (A, s, τ) be a dynamical system in a product MV-algebra $(A, \cdot)$, and X, Y be partitions in $(A, \cdot)$ such that $X \prec Y$. Then, for $\alpha > 1$, it holds $T_\alpha^s(\tau, X) \leq T_\alpha^s(\tau, Y)$.*

Proof. Let $X \prec Y$. By Propositions 2 and 6, we have $\vee_{k=0}^{n-1} \tau^k(X) \prec \vee_{k=0}^{n-1} \tau^k(Y)$, for $n = 1, 2, \ldots$. Therefore, by Theorem 2, we get:

$$T_\alpha^s(\vee_{k=0}^{n-1} \tau^k(X)) \leq T_\alpha^s(\vee_{k=0}^{n-1} \tau^k(Y)).$$

Consequently, dividing by n and letting $n \to \infty$, we get $T_\alpha^s(\tau, X) \leq T_\alpha^s(\tau, Y)$. $\square$

Definition 10. *The Tsallis entropy of a dynamical system* (A, s, τ) *in a product MV-algebra* $(A, \cdot)$ *is defined, for* $\alpha > 1$*, by:*

$$T_\alpha^s(\tau) = \sup\{T_\alpha^s(\tau, X) ; X \text{ is a partition in } (A, \cdot)\}.$$

Theorem 15. *Let* (A, s, τ) *be a dynamical system in a product MV-algebra* $(A, \cdot)$*. Then, for* $\alpha > 1$*, and every natural number* k*, it holds* $T_\alpha^s(\tau^k) = k \cdot T_\alpha^s(\tau)$*.*

Proof. Let X be a partition in $(A, \cdot)$. Then, for every natural number k, we have:

$$
\begin{aligned}
T_\alpha^s(\tau^k, \vee_{j=0}^{k-1} \tau^j(X)) &= \lim_{n \to \infty} \tfrac{1}{n} T_\alpha^s(\vee_{i=0}^{n-1} (\tau^k)^i (\vee_{j=0}^{k-1} \tau^j(X)) \\
&= \lim_{n \to \infty} \tfrac{1}{n} T_\alpha^s(\vee_{i=0}^{n-1} \vee_{j=0}^{k-1} \tau^{ki+j}(X)) = \lim_{n \to \infty} \tfrac{nk}{n} \tfrac{1}{nk} T_\alpha^s(\vee_{j=0}^{nk-1} \tau^j(X)) = k \cdot T_\alpha^s(\tau, X).
\end{aligned}
$$

Hence, we obtain:

$$
\begin{aligned}
k \cdot T_\alpha^s(\tau) &= k \cdot \sup\{T_\alpha^s(\tau, X) ; X \text{ is a partition in } (A, \cdot)\} \\
&= \sup\left\{ T_\alpha^s(\tau^k, \vee_{j=0}^{k-1} \tau^j(X)) ; X \text{ is a partition in } (A, \cdot) \right\} \\
&\leq \sup\left\{ T_\alpha^s(\tau^k, Y) ; Y \text{ is a partition in } (A, \cdot) \right\} = T_\alpha^s(\tau^k).
\end{aligned}
$$

On the other hand, by Proposition 1, we have $X \prec \vee_{j=0}^{k-1} \tau^j(X)$. Hence, by Theorem 14, we obtain:

$$T_\alpha^s(\tau^k, X) \leq T_\alpha^s(\tau^k, \vee_{j=0}^{k-1} \tau^j(X)) = k \cdot T_\alpha^s(\tau, X).$$

This implies that:

$$
\begin{aligned}
T_\alpha^s(\tau^k) &= \sup\left\{ T_\alpha^s(\tau^k, X) ; X \text{ is a partition in } (A, \cdot) \right\} \\
&\leq k \cdot \sup\{T_\alpha^s(\tau, X) ; X \text{ is a partition in } (A, \cdot)\} = k \cdot T_\alpha^s(\tau). \quad \square
\end{aligned}
$$

Definition 11. *Two product MV-algebra dynamical systems* (A_1, s_1, τ_1)*,* (A_2, s_2, τ_2) *are called isomorphic, if there exists some one-to-one and onto map* $\Phi : A_1 \to A_2$ *such that* $\Phi(u_1) = u_2$*, and, for every* $x, y \in A_1$*, the following conditions are satisfied:*

(i) $\Phi(x \cdot y) = \Phi(x) \cdot \Phi(y);$
(ii)   *if* $x + y \leq u_1$*, then* $\Phi(x + y) = \Phi(x) + \Phi(y);$
(iii) $s_2(\Phi(x)) = s_1(x);$
(iv) $\Phi(\tau_1(x)) = \tau_2(\Phi(x)).$

In this case, Φ *is said to be an isomorphism.*

Proposition 8. *Let* (A_1, s_1, τ_1)*,* (A_2, s_2, τ_2) *be isomorphic product MV-algebra dynamical systems, and* $\Phi : A_1 \to A_2$ *be an isomorphism between them. Then, for the inverse* $\Phi^{-1} : A_2 \to A_1$*, the following properties are satisfied:*

(i) $\Phi^{-1}(x \cdot y) = \Phi^{-1}(x) \cdot \Phi^{-1}(y), \text{ for every } x, y \in A_2;$
(ii)   *if* $x, y \in A_2$ *such that* $x + y \leq u_2$*, then* $\Phi^{-1}(x + y) = \Phi^{-1}(x) + \Phi^{-1}(y);$
(iii) $s_1(\Phi^{-1}(x)) = s_2(x), \text{ for every } x \in A_2;$
(iv) $\Phi^{-1}(\tau_2(x)) = \tau_1(\Phi^{-1}(x)), \text{ for every } x \in A_2.$

Proof. The map $\Phi : A_1 \to A_2$ is bijective, therefore, for every $x, y \in A_2$, there exist $x', y' \in A_1$ such that $\Phi^{-1}(x) = x'$, and $\Phi^{-1}(y) = y'$.

(i)   Let $x, y \in A_2$. Then we get:

$$\Phi^{-1}(x \cdot y) = \Phi^{-1}(\Phi(x') \cdot \Phi(y')) = \Phi^{-1}(\Phi(x' \cdot y')) = x' \cdot y' = \Phi^{-1}(x) \cdot \Phi^{-1}(y).$$

(ii)   Let $x, y \in A_2$ such that $x + y \leq u_2$. Then $x' + y' \leq u_1$, and, therefore, we have:

$$\Phi^{-1}(x + y) = \Phi^{-1}(\Phi(x') + \Phi(y')) = \Phi^{-1}(\Phi(x' + y')) = x' + y' = \Phi^{-1}(x) + \Phi^{-1}(y).$$

(iii)   Let $x \in A_2$. Then $s_2(x) = s_2(\Phi(x')) = s_1(x') = s_1(\Phi^{-1}(x))$.

(iv)   Let $x \in A_2$. Then $\Phi^{-1}(\tau_2(x)) = \Phi^{-1}(\tau_2(\Phi(x'))) = \Phi^{-1}(\Phi(\tau_1(x'))) = \tau_1(x') = \tau_1(\Phi^{-1}(x))$. $\square$

Theorem 16.   *Let* (A_1, s_1, τ_1), (A_2, s_2, τ_2) *be isomorphic product MV-algebra dynamical systems, and* $\alpha > 1$. *Then:*

$$T_\alpha^{s_1}(\tau_1) = T_\alpha^{s_2}(\tau_2).$$

Proof.   Let $\Phi : A_1 \to A_2$ be an isomorphism between dynamical systems (A_1, s_1, τ_1), (A_2, s_2, τ_2). Consider a partition $X = (x_1, x_2, \ldots, x_k)$ in a product MV-algebra $(A_1, \cdot)$. Then $x_1 + x_2 + \ldots + x_k = u_1$, and therefore, by the condition (ii) of Definition 11, it holds $\Phi(x_1) + \Phi(x_2) + \ldots + \Phi(x_k) = \Phi(x_1 + x_2 + \ldots + x_k) = \Phi(u_1) = u_2$. This means that the k-tuple $\Phi(X) = (\Phi(x_1), \Phi(x_2), \ldots, \Phi(x_k))$ is a partition in a product MV-algebra $(A_2, \cdot)$. Moreover, according to the condition (iii) of Definition 11, we have:

$$T_\alpha^{s_2}(\Phi(X)) = \sum_{i=1}^{k} l_\alpha(s_2(\Phi(x_i))) = \sum_{i=1}^{k} l_\alpha(s_1(x_i)) = T_\alpha^{s_1}(X).$$

Hence, using the conditions (iv), and (i) of Definition 11, we get:

$$T_\alpha^{s_2}(\vee_{k=0}^{n-1} \tau_2^k(\Phi(X))) = T_\alpha^{s_2}(\vee_{k=0}^{n-1} \Phi(\tau_1^k(X)))$$
$$= T_\alpha^{s_2}(\Phi(\vee_{k=0}^{n-1} \tau_1^k(X))) = T_\alpha^{s_1}(\vee_{k=0}^{n-1} \tau_1^k(X)).$$

Therefore, dividing by n and letting $n \to \infty$, we obtain:

$$T_\alpha^{s_2}(\tau_2, \Phi(X)) = \lim_{n \to \infty} \frac{1}{n} T_\alpha^{s_2}(\vee_{k=0}^{n-1} \tau_2^k(\Phi(X))) = \lim_{n \to \infty} \frac{1}{n} T_\alpha^{s_1}(\vee_{k=0}^{n-1} \tau_1^k(X)) = T_\alpha^{s_1}(\tau_1, X).$$

This implies that:

$$\{T_\alpha^{s_1}(\tau_1, X) ; X \text{ is a partition in } (A_1, \cdot)\} \subset \{T_\alpha^{s_2}(\tau_2, Y) ; Y \text{ is a partition in } (A_2, \cdot)\},$$

and consequently:

$$T_\alpha^{s_1}(\tau_1) = \sup\{T_\alpha^{s_1}(\tau_1, X) ; X \text{ is a partition in } (A_1, \cdot)\}$$
$$\leq \sup\{T_\alpha^{s_2}(\tau_2, Y) ; Y \text{ is a partition in } (A_2, \cdot)\} = T_\alpha^{s_2}(\tau_2).$$

The converse $T_\alpha^{s_2}(\tau_2) \leq T_\alpha^{s_1}(\tau_1)$ can be obtained in a similar way; according to Proposition 8, it suffices to consider the inverse $\Phi^{-1} : A_2 \to A_1$. $\square$

Remark 5. *It trivially follows from Theorem 16 that if* $T_\alpha^{s_1}(\tau_1) \neq T_\alpha^{s_2}(\tau_2)$, *then the corresponding dynamical systems* (A_1, s_1, τ_1), (A_2, s_2, τ_2) *are not isomorphic. This means that some product MV-algebra dynamical systems can be distinguished due to their different Tsallis entropies.*

6. Conclusions

In this article we dealt with the mathematical modelling of Tsallis entropy in product MV-algebras. Our results are given in Sections 3–5. In Section 3 we have introduced the notion of Tsallis entropy $T_\alpha^s(X)$ of a partition X in a product MV-algebra $(A, \cdot)$, and we examined properties of this entropy measure. In Section 4 we have defined and studied the conditional Tsallis entropy of partitions in this algebraic structure. It has been shown that the proposed concepts are consistent, in the case of the limit of $\alpha \to 1$, with the Shannon entropy expressed in nats, defined and studied in Reference [31]. Moreover, putting $\alpha = 2$ in the proposed definitions, we obtain the logical entropy of partitions in a product MV-algebra defined and studied in Reference [8].

Section 5 was devoted to the mathematical modelling of Tsallis entropy in product MV-algebra dynamical systems. From Example 8 it follows that the notion of product MV-algebra dynamical system is a generalization of the concept of classical dynamical system. We have shown that the Tsallis entropy is invariant under isomorphism of product MV-algebra dynamical systems.

In the proofs we used L'Hôpital's rule and the known Jensen inequality. To illustrate the results, we have provided several numerical examples. In Example 2, we have mentioned that the full tribe of fuzzy sets is a special case of product MV-algebras; hence, all the results of this article can be directly applied to this family of fuzzy sets. We remind that a fuzzy subset of a non-empty set Ω is any mapping $f : \Omega \to [0,1]$, where the value $f(\omega)$ is interpreted as the degree of belonging of element ω of Ω to the fuzzy set f (cf. [16]). In Reference [56], Atanassov has generalized the Zadeh fuzzy set theory by introducing the idea of an intuitionistic fuzzy set (IF-set), a set that has the degree of belonging as well as the degree of non-belonging with each of its elements. From the point of view of application, it should be noted that for a given class $\mathcal{F}$ of IF-sets can be created an MV-algebra $\mathcal{A}$ such that $\mathcal{F}$ can be inserted to $\mathcal{A}$. Also the operation of product on $\mathcal{F}$ can be defined by such a way that the corresponding MV-algebra is a product MV-algebra. Therefore, the presented results are also applicable to the case of IF-sets.

Author Contributions: Conceptualization, D.M.; Formal analysis, D.M. and B.R.; Investigation, D.M. and Beloslav Riečan; Writing original draft, D.M.; Writing review & editing, B.R. Both authors have read and approved the final manuscript.

Acknowledgments: The authors thank the editor and the referees for their valuable comments and suggestions.

Conflicts of Interest: The authors declare no conflict of interest.

References

1. Shannon, C.E. A mathematical theory of communication. *Bell Syst. Tech. J.* **1948**, *27*, 379–423. [CrossRef]
2. Gray, R.M. *Entropy and Information Theory*; Springer: Berlin/Heidelberg, Germany, 2009.
3. Kolmogorov, A.N. New metric invariant of transitive dynamical systems and automorphisms of Lebesgue spaces. *Dokl. Russ. Acad. Sci.* **1958**, *119*, 861–864.
4. Sinai, Y.G. On the notion of entropy of a dynamical system. *Dokl. Russ. Acad. Sci.* **1959**, *124*, 768–771.
5. Markechová, D.; Ebrahimzadeh, A.; Eslami Giski, Z. Logical entropy of dynamical systems. *Adv. Differ. Equ.* **2018**, *2018*, 70. [CrossRef]
6. Ellerman, D. Logical information theory: New foundations for information theory. *Log. J. IGPL* **2017**, *25*, 806–835. [CrossRef]
7. Markechová, D.; Riečan, B. Logical entropy of fuzzy dynamical systems. *Entropy* **2016**, *18*, 157. [CrossRef]
8. Markechová, D.; Mosapour, B.; Ebrahimzadeh, A. Logical Divergence, Logical Entropy, and Logical Mutual Information in Product MV-Algebras. *Entropy* **2018**, *20*, 129. [CrossRef]
9. Mohammadi, U. The concept of logical entropy on D-posets. *J. Algebraic Struct. Appl.* **2016**, *1*, 53–61.
10. Ebrahimzadeh, A. Logical entropy of quantum dynamical systems. *Open Phys.* **2016**, *14*, 1–5. [CrossRef]
11. Ebrahimzadeh, A. Quantum conditional logical entropy of dynamical systems. *Ital. J. Pure Appl. Math.* **2016**, *36*, 879–886.
12. Markechová, D.; Riečan, B. Logical entropy and logical mutual information of experiments in the intuitionistic fuzzy case. *Entropy* **2017**, *19*, 429. [CrossRef]
13. Ebrahimzadeh, A.; Eslami Giski, Z.; Markechová, D. Logical entropy of dynamical systems—A general model. *Mathematics* **2017**, *5*, 4. [CrossRef]
14. Ebrahimzadeh, A.; Jamalzadeh, J. Conditional logical entropy of fuzzy σ-algebras. *J. Intell. Fuzzy Syst.* **2017**, *33*, 1019–1026. [CrossRef]

15. Eslami Giski, Z.; Ebrahimzadeh, A. An introduction of logical entropy on sequential effect algebra. *Indag. Math.* **2017**, *28*, 928–937. [CrossRef]
16. Zadeh, L.A. Fuzzy Sets. *Inf. Control* **1965**, *8*, 338–358. [CrossRef]
17. Zadeh, L.A. Probability measures of fuzzy events. *J. Math. Anal. Appl.* **1968**, *23*, 421–427. [CrossRef]
18. Markechová, D. The entropy of fuzzy dynamical systems and generators. *Fuzzy Sets Syst.* **1992**, *48*, 351–363. [CrossRef]
19. Dumitrescu, D. Entropy of a fuzzy dynamical system. *Fuzzy Sets Syst.* **1995**, *70*, 45–57. [CrossRef]
20. Riečan, B.; Markechová, D. The entropy of fuzzy dynamical systems, general scheme, and generators. *Fuzzy Sets Syst.* **1998**, *96*, 191–199. [CrossRef]
21. Markechová, D.; Riečan, B. Entropy of fuzzy partitions and entropy of fuzzy dynamical systems. *Entropy* **2016**, *18*, 19. [CrossRef]
22. Chang, C.C. Algebraic analysis of many valued logics. *Trans. Am. Math. Soc.* **1958**, *88*, 467–490. [CrossRef]
23. Mundici, D. Interpretation of AFC*-algebras in Łukasiewicz sentential calculus. *J. Funct. Anal.* **1986**, *56*, 889–894.
24. Mundici, D. *Advanced Łukasiewicz Calculus and MV-Algebras*; Springer: Dordrecht, The Netherlands, 2011.
25. Di Nola, A.; Dvurečenskij, A.; Hyčko, M.; Manara, C. Entropy on Effect Algebras with the Riesz Decomposition Property II: MV-Algebras. *Kybernetika* **2005**, *41*, 161–176.
26. Riečan, B. Kolmogorov–Sinaj entropy on MV-algebras. *Int. J. Theor. Phys.* **2005**, *44*, 1041–1052. [CrossRef]
27. Riečan, B.; Mundici, D. Probability on MV-algebras. In *Handbook of Measure Theory*; Pap, E., Ed.; Elsevier: Amsterdam, The Netherlands, 2002; pp. 869–910.
28. Riečan, B. On the product MV-algebras. *Tatra Mt. Math.* **1999**, *16*, 143–149.
29. Montagna, F. An algebraic approach to propositional fuzzy logic. *J. Log. Lang. Inf.* **2000**, *9*, 91–124. [CrossRef]
30. Riečan, B.; Neubrunn, T. *Integral, Measure and Ordering*; Springer: Dordrecht, The Netherlands, 1997.
31. Petrovičová, J. On the entropy of partitions in product MV-algebras. *Soft Comput.* **2000**, *4*, 41–44. [CrossRef]
32. Petrovičová, J. On the entropy of dynamical systems in product MV-algebras. *Fuzzy Sets Syst.* **2001**, *121*, 347–351. [CrossRef]
33. Tsallis, C. Possible generalization of Boltzmann–Gibbs statistics. *J. Stat. Phys.* **1988**, *52*, 479–487. [CrossRef]
34. Havrda, J.; Charvát, F. Quantification methods of classification processes: Concept of structural alpha-entropy. *Kybernetika* **1967**, *3*, 30–35.
35. Tsallis, C. *Introduction to Nonextensive Statistical Mechanics: Approaching a Complex World*; Springer: New York, NY, USA, 2009.
36. Hanel, R.; Thurner, S. Generalized Boltzmann factors and the maximum entropy principle: Entropies for complex systems. *Phys. A Stat. Mech Appl.* **2007**, *380*, 109–114. [CrossRef]
37. Almeida, M.P. Generalized entropies from first principles. *Phys. A Stat. Mech Appl.* **2001**, *300*, 424–432. [CrossRef]
38. Kaniadakis, G. Statistical mechanics in the context of special relativity. *Phys. Rev. E Stat. Nonlin. Soft Matter Phys.* **2002**, *66*, 056125. [CrossRef] [PubMed]
39. Naudts, J. Deformed exponentials and logarithms in generalized thermostatistics. *Phys. A Stat. Mech. Appl.* **2002**, *316*, 323–334. [CrossRef]
40. Tsallis, C. Generalized entropy-based criterion for consistent testing. *Phys. Rev. E* **1998**, *58*, 1442–1445. [CrossRef]
41. Alemany, P.A.; Zanette, D.H. Fractal random walks from a variational formalism for Tsallis entropies. *Phys. Rev. E* **1994**, *49*, R956–R958. [CrossRef]
42. Tsallis, C. Nonextensive thermostatistics and fractals. *Fractals* **1995**, *3*, 541. [CrossRef]
43. Tsallis, C.; Anteneodo, C.; Borland, L.; Osorio, R. Nonextensive Statistical mechanics and economics. *Phys. A Stat. Mech. Appl.* **2003**, *324*, 89–100. [CrossRef]
44. Borland, L. Long-range memory and nonextensivity in financial markets. *Europhys. News* **2005**, *36*, 228–231. [CrossRef]
45. Pérez, D.G.; Zunino, L.; Martín, M.T.; Pérez, D.G.; Zunino, L.; Martín, M.T.; Garavaglia, M.; Plastino, A.; Rosso, O.A. Model-free stochastic processes studied with q-wavelet-based in formational tools. *Phys. Lett. A* **2007**, *364*, 259–266. [CrossRef]
46. Huang, H.; Xie, H.; Wang, Z. The analysis of VF and VT with wavelet-based Tsallis information measure. *Phys. Lett. A* **2005**, *336*, 180–187. [CrossRef]

47. Tong, S.; Bezerianos, A.; Paul, J.; Zhu, Y.; Thakor, N. Nonextensive entropy measure of EEG following brain injury from cardiac arrest. *Phys. A Stat. Mech. Appl.* **2002**, *305*, 619–628. [CrossRef]
48. Rosso, O.A.; Martín, M.T.; Plastino, A. Brain electrical activity analysis using wavelet-based informational tools (II): Tsallis non-extensivity and complexity measures. *Phys. A Stat. Mech. Appl.* **2003**, *320*, 497–511. [CrossRef]
49. Kumar, V. *Kapur's and Tsalli's Entropies: A Communication System Perspective*; LAP LAMBERT Academic Publishing: Saarbrucken, Germany, 2015.
50. Gluschankof, D. Cyclic ordered groups and MV-algebras. *Czechoslovak Math. J.* **1993**, *43*, 249–263.
51. Cattaneo, G.; Lombardo, F. Independent axiomatization for MV-algebras. *Tatra Mt. Math.* **1998**, *15*, 227–232.
52. Riečan, B. Analysis of Fuzzy Logic Models. In *Intelligent Systems*; Koleshko, V.M., Ed.; InTech: Rijeka, Croatia, 2012; pp. 219–244.
53. Anderson, M.; Feil, T. *Lattice Ordered Groups*; Kluwer: Dordrecht, The Netherlands, 1988.
54. Riečan, B. On the probability theory on product MV-algebras. *Soft Comput.* **2000**, *4*, 49–57. [CrossRef]
55. Walters, P. *An Introduction to Ergodic Theory*; Springer: New York, NY, USA, 1982.
56. Atanassov, K. Intuitionistic fuzzy sets. *Fuzzy Sets Syst.* **1986**, *20*, 87–96. [CrossRef]

Article

A Novel Image Encryption Scheme Based on Self-Synchronous Chaotic Stream Cipher and Wavelet Transform

Chunlei Fan and Qun Ding *

Electrical Engineering College, Heilongjiang University, Harbin 150080, China; 1172053@s.hlju.edu.cn
* Correspondence: 1984008@hlju.edu.cn; Tel.: +86-0451-8660-8504

Received: 21 May 2018; Accepted: 5 June 2018; Published: 6 June 2018

Abstract: In this paper, a novel image encryption scheme is proposed for the secure transmission of image data. A self-synchronous chaotic stream cipher is designed with the purpose of resisting active attack and ensures the limited error propagation of image data. Two-dimensional discrete wavelet transform and Arnold mapping are used to scramble the pixel value of the original image. A four-dimensional hyperchaotic system with four positive Lyapunov exponents serve as the chaotic sequence generator of the self-synchronous stream cipher in order to enhance the security and complexity of the image encryption system. Finally, the simulation experiment results show that this image encryption scheme is both reliable and secure.

Keywords: hyperchaotic system; self-synchronous stream cipher; permutation entropy; image encryption; wavelet transform

---

1. Introduction

With the rapid development of social networking, cloud computing, and mobile network communication technology, the problem of secure storage and real-time transmission of image data is increasingly important. Encryption and digital watermarking technology play an important role in guaranteeing the security of multimedia data [1]. However, because of the high correlation and redundancy of adjacent pixels of the digital image, some international standard encryption algorithms are not suitable for image encryption, including 3DES (Triple Data Encryption Algorithm), IDEA (International Data Encryption Algorithm), and AES (Advanced Encryption Standard), etc. On the other hand, the chaotic nonlinear dynamic system has some good characteristics, such as positive Lyapunov exponents, ergodicity, sensitivity to initial conditions, topological transitivity, and unpredictability [2–5], and was widely applied in the field of cryptography and secret communication. In recent years, in order to better solve the security transmission of digital images, some scholars have put forward a series of image security encryption schemes based on the chaotic system and the inherent characteristics of digital images [6–9]. For example, Ping et al. [10] proposed a permutation-substitution image encryption scheme with the Henon map, which can resist a chosen-plaintext attack and known-plaintext attack. Ye et al. [11] put forward an efficient symmetric image encryption algorithm based on an intertwining Logistic map. Haroun [12] came up with a real-time image encryption scheme using a low-complexity discrete 3D dual chaotic cipher.

However, these image encryption schemes generally use low-dimensional chaotic systems or high-dimensional chaotic systems with only one positive Lyapunov exponent such as Logistic, Tent, Henon, and Lorenz, etc. Compared with the high-dimensional hyperchaotic system with more than two positive Lyapunov exponents, the complexity of nonlinear dynamic characteristics of the above chaotic systems are lower. Additionally, because of the influence of the calculation precision and the quantization method, the chaotic binary sequences generated by the low-dimensional chaotic systems emerge with short periodic phenomena [13,14], which will seriously affect the security of image encryption. Furthermore, the above image encryption schemes usually adopt the synchronous sequence cipher based on the chaotic binary sequences [15,16]. In synchronous stream ciphers, the key stream is independent of plaintext or ciphertext. In the process of communication, the sender and receiver must keep accurate synchronization. If the synchronization mechanism is broken by active attack, the receiver will not be able to decrypt the ciphertext correctly. For instance, if an attacker inserts or removes a certain number of bits ciphertext, it will immediately destroy the synchronization mechanism of the synchronous sequence cipher. Therefore, this encryption method cannot resist active attack [17]. On the basis of the above image encryption problem, we proposed a novel image encryption scheme based on self-synchronous chaotic stream cipher and wavelet transform. Firstly, a two-dimensional discrete wavelet transform is used to convert the original image from the spatial domain to the frequency domain with the purpose of strengthening the difficulty of cracking. Secondly, the pixel value of the image is scrambled by Arnold mapping. Finally, the scrambled image is encrypted by self-synchronous chaotic stream cipher. This algorithm uses a four-dimensional hyperchaotic system with four positive Lyapunov exponents and a self-synchronous stream cipher mechanism. The generation of the key stream of the self-synchronous stream cipher is not independent of the plaintext and ciphertext stream but is related to the seed key and n-bits ciphertext that have been generated before. In the process of ciphertext transmission, the 1-bit ciphertext error will only affect the correct decryption of the n-bits ciphertext in the back. The decryption process returns to normal after this time. Therefore, this scheme cannot only resist active attack but also ensures the limited error propagation of image data. The experimental results show that the encryption scheme has good security.

The rest of this paper is organized as follows: Section 2 introduces a four-dimensional hyperchaotic system and the design scheme of the self-synchronous chaotic stream cipher. Furthermore, the performance of the discrete chaotic sequence was analyzed by multi-scale permutation entropy and NIST-800-22 test. In Section 3, a novel image encryption scheme is proposed and a detailed security analysis is carried out with histogram and information entropy analyses, etc. Section 4 summarizes the conclusion of this paper.

2. Design and Implementation of Self-Synchronous Chaotic Stream Cipher

2.1. The Description of Four-Dimensional Discrete Chaotic System

In this section, a four-dimensional chaotic system is constructed through the Chen–Lai algorithm [18,19]. The discrete dynamic equations of the system can be expressed as the follows:

$$\begin{pmatrix} x_1(k+1) \\ x_2(k+1) \\ x_3(k+1) \\ x_4(k+1) \end{pmatrix} = A \begin{pmatrix} x_1(k) \\ x_2(k) \\ x_3(k) \\ x_4(k) \end{pmatrix} + (\|A\|_2 + e^c) \begin{pmatrix} x_1(k) \\ x_2(k) \\ x_3(k) \\ x_4(k) \end{pmatrix} (\mathrm{mod}1) \tag{1}$$

where $\|\cdot\|_2$ and e represent Euclidean norm and mathematical constant, respectively. The mod is the module operations, and c is control parameter. Furthermore, matrix A is given as follows:

$$A = \begin{pmatrix} A_{11} & A_{12} & A_{13} & A_{14} \\ A_{21} & A_{22} & A_{23} & A_{24} \\ A_{31} & A_{32} & A_{33} & A_{34} \\ A_{41} & A_{42} & A_{43} & A_{44} \end{pmatrix} = \begin{pmatrix} 0.7 & 0.4 & 0.1 & 0.2 \\ 0.2 & -0.5 & 0.1 & 0 \\ 0 & -1/3 & 0.1 & 0 \\ 0 & -1/4 & 0.3 & 0.6 \end{pmatrix}. \tag{2}$$

When $c = 3$, the Lyapunov exponents of the system are given by $LE_1 = 3.0128$, $LE_2 = 3.0454$, $LE_3 = 3.0717$, and $LE_4 = 3.0799$. The number of positive Lyapunov exponents are more than two. Thus, the system is a four-dimensional hyperchaotic system. The chaotic time series of the four-dimensional hyperchaotic system are shown in Figure 1.

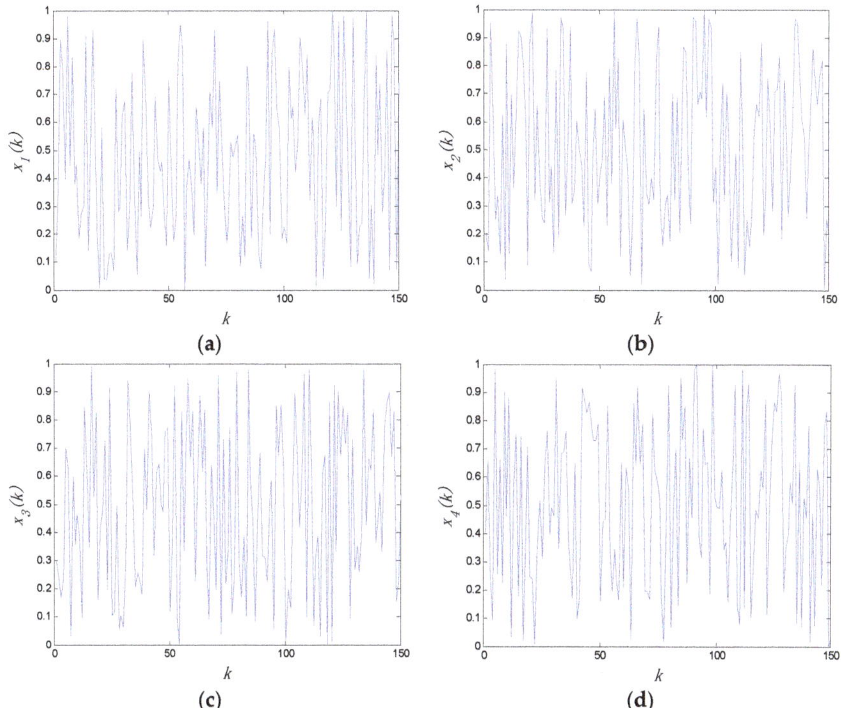

Figure 1. The chaotic time series of the four-dimensional hyperchaotic system with: (**a**) $x_1(k)$; (**b**) $x_2(k)$; (**c**) $x_3(k)$; (**d**) $x_4(k)$.

2.2. Quantization and Performance Analysis of Discrete Chaotic Sequences

2.2.1. Binary Quantization Method

For the above chaotic system, $x_j(k) \in (0,1)$ with $j = 1, 2, 3, 4$. In this paper, we adopt the binary quantization method to quantize discrete chaotic real value sequences. The corresponding quantization method is defined as follows:

$$Q_j(k) = \left\{ \begin{array}{ll} 0 & x_j(k) < t_d \\ 1 & x_j(k) \geq t_d \end{array} \right. \quad j = 1, 2, 3, 4 \tag{3}$$

where $Q_j(k)$ is the quantized chaotic binary sequence, and t_d represents the quantization threshold with $t_d = 0.5$.

2.2.2. Multi-Scale Permutation Entropy Analysis

Multi-scale permutation entropy (MPE) [20,21] has the advantages of high robustness and fast computational speed. It is widely applied in the measurement of binary sequence complexity and nonlinear system analysis. In this section, we perform a multi-scale permutation entropy analysis for the above chaotic binary sequence. The parameters of MPE have embedding dimension m, delay factor τ, and scale factor s. For the choice of the parameter values of the multi-scale permutation entropy with the purpose of calculating the complexity of chaotic binary sequences, Sun et al. [22] and Xu et al. [23] give the recommended parameter range in order to obtain more accurate entropy values. On the basis of the theoretical research of the above references, in this experiment, we set $m = 3, \tau = 2$, and $s \in [3,7]$, respectively. The experimental results are shown in Table 1. As can be seen from Table 1, all MPE values of chaotic binary sequences are more than 0.9 and display good sequence complexity.

Table 1. The multi-scale permutation entropy (MPE) value of chaotic binary sequences with $Q_1(k)$, $Q_2(k)$, $Q_3(k)$, and $Q_4(k)$.

Scale Factor S	$Q_1(k)$	$Q_2(k)$	$Q_3(k)$	$Q_4(k)$
3	0.9201	0.9260	0.9177	0.9132
4	0.9366	0.9382	0.9449	0.9410
5	0.9548	0.9488	0.9567	0.9377
6	0.9590	0.9572	0.9585	0.9526
7	0.9552	0.9533	0.9704	0.9553

2.2.3. NIST-800-22 Test

NIST-800-22 is a statistical test suite for random and pseudorandom number generators for cryptographic applications. This test standard was enacted by the National Institute of Standards and Technology (NIST). The test statistic is used to calculate a p-value that summarizes the strength of the evidence against the null hypothesis. On the basis of the results of NIST test, we can judge whether or not this chaotic binary sequence is suitable for a cryptographic algorithm. NIST-800-22 is made up of 16 test methods, including the longest run test, cumulative sums, and the linear complexity test, etc. For these tests, each p-value is the probability that a perfect random number generator would have produced a sequence less random than the sequence that was tested, given the kind of non-randomness assessed by the test. A significance level (α) can be chosen for the tests. If $p - \text{value} \geq \alpha$, then the null hypothesis is accepted; i.e., the sequence appears to be random. If $p - \text{value} < \alpha$, then the null hypothesis is rejected; i.e., the sequence appears to be non-random. Typically, α is chosen in the range $[0.001, 0.01]$. Common values of α in cryptography are about 0.01 based on the NIST-800-22 test standard [24]. The experimental results of NIST-800-22 test are shown in Table 2. Table 2 shows that the chaotic binary sequences $Q_1(k)$, $Q_2(k)$, $Q_3(k)$, and $Q_4(k)$ passed all the tests. These sequences show good randomness and meet the requirements of the stream cipher.

Table 2. NIST-800-22 test of chaotic binary sequences.

Test Item	$Q_1(k)$	$Q_2(k)$	$Q_3(k)$	$Q_4(k)$	Result
	$p - \text{Value}$	$p - \text{Value}$	$p - \text{Value}$	$p - \text{Value}$	
Approximate Entropy	0.026853	0.013829	0.068205	0.034937	Success
Block Frequency	0.058378	0.870831	0.724584	0.297646	Success
Cumulative Sums	0.459642	0.069717	0.963210	0.328997	Success
FFT	0.358795	0.919848	0.081236	0.713570	Success
Frequency	0.435391	0.447255	0.888660	0.193601	Success
Linear Complexity	0.186537	0.203633	0.569565	0.232544	Success
Longest Run	0.359643	0.087189	0.789913	0.250387	Success
Non-Overlapping Template	0.348045	0.680967	0.106169	0.068529	Success
Overlapping Template	0.512834	0.063236	0.020689	0.490518	Success
Random Excursions	0.319514	0.181174	0.524622	0.304589	Success
Random Excursions Variant	0.579380	0.177934	0.108254	0.659874	Success
Rank	0.949536	0.648387	0.862457	0.648387	Success
Runs	0.340097	0.086469	0.041369	0.027231	Success
Serial Test-1	0.407933	0.213432	0.648688	0.814738	Success
Serial Test-2	0.462490	0.880617	0.584615	0.512974	Success
Maurer's Universal	0.026152	0.538143	0.142680	0.600293	Success

2.3. The Design of Self-Synchronous Chaotic Stream Cipher

The self-synchronous stream cipher is also known as the asynchronous stream cipher. The generation of the key stream of self-synchronous stream cipher is not independent of the plaintext and ciphertext stream but is related to the seed key and some ciphertext that has been generated before.

It has the advantages of limited error propagation, self-synchronous and ciphertext statistical diffusion. In this section, the encryption and decryption block diagram of the self-synchronous stream cipher based on the four-dimensional hyperchaotic system is shown in Figure 2.

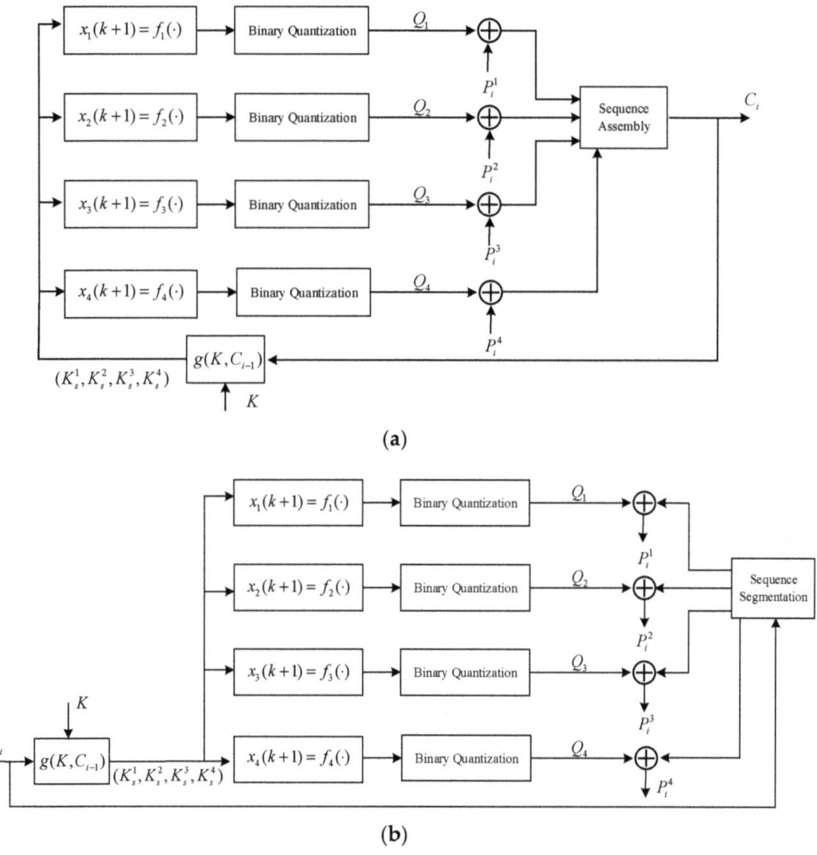

Figure 2. The self-synchronous chaotic stream cipher with: (**a**) decryption block diagram; (**b**) decryption block diagram.

Where K, $(K_s^1, K_s^2, K_s^3, K_s^4)$, $\oplus$ and function $g(\cdot)$ represent the seed key, subkey, exclusive OR (XOR) operational character and subkey generation function, respectively. C_{i-1} and C_i are the ciphertext stream generated at two adjacent moments. Q_1, Q_2, Q_3, and Q_4 are the quantized chaotic binary sequence. P_i^1, P_i^2, P_i^3, and P_i^4 are the adjacent plaintext stream with the purpose of parallel encryption. According to Equations (1) and (2), $f_1(\cdot)$, $f_2(\cdot)$, $f_3(\cdot)$, and $f_4(\cdot)$ can be given as Equation (4).

$$\begin{cases} x_1(k+1) = f_1(\cdot) = A_{11}x_1(k) + A_{12}x_2(k) + A_{13}x_3(k) + A_{14}x_4(k) + (\|A\|_2 + e^c)x_1(k)(\mathrm{mod}1) \\ x_2(k+1) = f_2(\cdot) = A_{21}x_1(k) + A_{22}x_2(k) + A_{23}x_3(k) + A_{24}x_4(k) + (\|A\|_2 + e^c)x_2(k)(\mathrm{mod}1) \\ x_3(k+1) = f_3(\cdot) = A_{31}x_1(k) + A_{32}x_2(k) + A_{33}x_3(k) + A_{34}x_4(k) + (\|A\|_2 + e^c)x_3(k)(\mathrm{mod}1) \\ x_4(k+1) = f_4(\cdot) = A_{41}x_1(k) + A_{42}x_2(k) + A_{43}x_3(k) + A_{44}x_4(k) + (\|A\|_2 + e^c)x_4(k)(\mathrm{mod}1) \end{cases} \quad (4)$$

The encryption process of the self-synchronous chaotic stream cipher can be described as follows:

1. The subkey $(K_s^1, K_s^2, K_s^3, K_s^4)$ is generated by function $g(\cdot)$ and $K_s^j \in (0,1)$ with $j = 1,2,3,4$. Where K and C_{i-1} are the 0-1 binary sequence with length of 32 bits. The function $g(\cdot)$ is given as follows:

$$
\left\{
\begin{array}{l}
K = (k_1, k_2, \cdots, k_{32}) \\
C_{i-1} = (c_{i-1}^1, c_{i-1}^2, \cdots, c_{i-1}^{32}) \\
K_s^1 = g(K, C_{i-1}) = \sum_{v=1}^{8} (k_{4v-3} \oplus c_{i-1}^{4v-3}) 2^{-v} \\
K_s^2 = g(K, C_{i-1}) = \sum_{v=1}^{8} (k_{4v-2} \oplus c_{i-1}^{4v-2}) 2^{-v} \\
K_s^3 = g(K, C_{i-1}) = \sum_{v=1}^{8} (k_{4v-1} \oplus c_{i-1}^{4v-1}) 2^{-v} \\
K_s^4 = g(K, C_{i-1}) = \sum_{v=1}^{8} (k_{4v} \oplus c_{i-1}^{4v}) 2^{-v}
\end{array}
\right.
\tag{5}
$$

2. The generated subkey $(K_s^1, K_s^2, K_s^3, K_s^4)$ is used as the state variable $(x_1(k), x_2(k), x_3(k), x_4(k))$ of the four-dimensional hyperchaotic system. The key stream Q_1, Q_2, Q_3 and, Q_4 with length of 8 bits are generated by hyperchaotic Equation (4) and a binary quantization operation with the purpose of parallel encryption.

3. Ciphertext C_i is generated by the Equation (6). At the same time, the C_i will feedback to the function $g(\cdot)$ with the purpose of generating the next round subkey $(K_s^1, K_s^2, K_s^3, K_s^4)$. Where $\|$ is sequence assembly operation.

$$
\left\{
\begin{array}{l}
Q_1 \oplus P_i^1 = c_i^1, c_i^2, \cdots, c_i^8 \\
Q_2 \oplus P_i^2 = c_i^9, c_i^{10}, \cdots, c_i^{16} \\
Q_3 \oplus P_i^3 = c_i^{17}, c_i^{18}, \cdots, c_i^{24} \\
Q_4 \oplus P_i^4 = c_i^{25}, c_i^{26}, \cdots, c_i^{32} \\
C_i = (Q_1 \oplus P_i^1) \| (Q_2 \oplus P_i^2) \| (Q_3 \oplus P_i^3) \| (Q_4 \oplus P_i^4) = c_i^1, c_i^2, \cdots, c_i^{32}
\end{array}
\right.
\tag{6}
$$

The decryption process of the self-synchronous chaotic stream cipher is similar to the encryption process, which is not repeated here.

3. A Novel Image Encryption Scheme Based on Self-Synchronous Chaotic Stream Cipher

3.1. The Description of the Image Encryption Scheme

In this section, a novel image encryption scheme is proposed based on self-synchronous chaotic stream cipher, Arnold mapping, and two-dimensional discrete wavelet transform (DWT). The encryption process of the proposed scheme is shown in Figure 3. Firstly, the spatial domain of the image is transformed into the frequency domain by two-layer DWT. Secondly, Arnold mapping is implemented with the purpose of obtaining good diffusion effectiveness. Finally, on the basis of self-synchronous chaotic stream cipher, the scrambled image is encrypted to ensure its security.

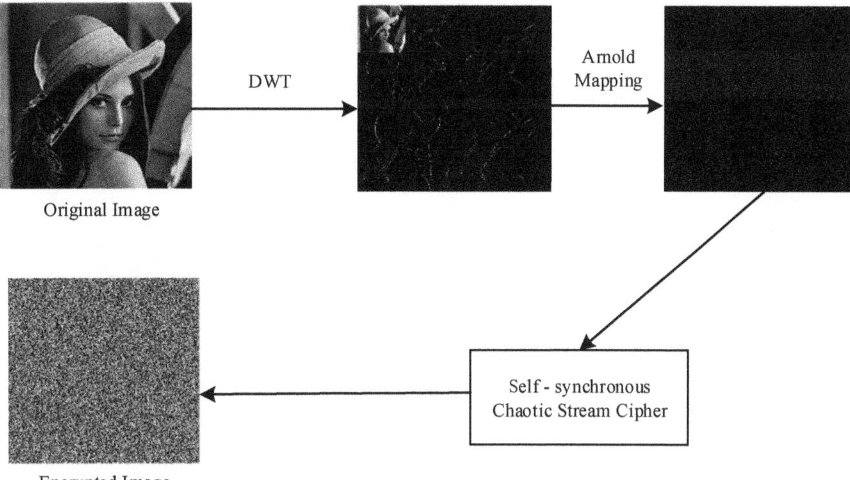

Figure 3. The encryption process of the proposed scheme.

3.1.1. Discrete Wavelet Transform

The DWT plays an important role in image compression and image information processing. The decomposition process of signal S_0 by the two-dimensional DWT is shown in Figure 4. A two-dimensional matrix S_0 can be decomposed into four groups of coefficients $[cA, cD^{(h)}, cD^{(v)}, cD^{(d)}]$. Where cA, $cD^{(h)}$, $cD^{(v)}$, and $cD^{(d)}$ represent approximate coefficient (low frequency component), horizontal detail coefficient, vertical detail coefficient, and diagonal detail coefficients, respectively. Furthermore, the approximate coefficient cA can continue to be decomposed by the same method. On the basis of the above image encryption scheme, in order to obtain the transformation coefficient, the pixel value of the grayscale image is processed by two-dimensional discrete wavelet transform. Thereafter, the spatial domain of the digital image is transformed into the frequency domain with the purpose of enhancing the pixel scrambling effect.

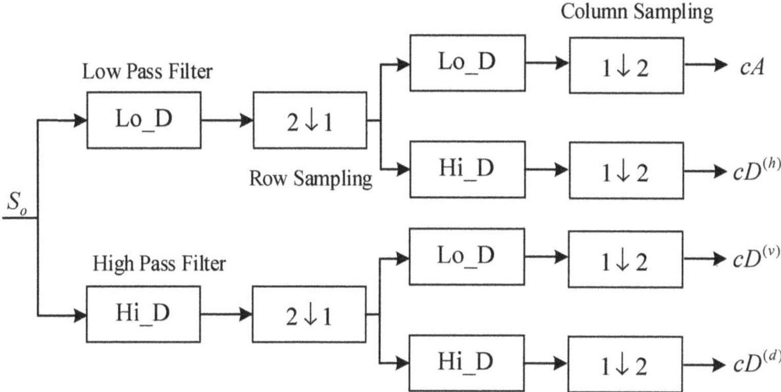

Figure 4. The principle block diagram of the two-dimensional discrete wavelet transform.

3.1.2. Arnold Mapping

Arnold mapping [25], also known as cat mapping, is a chaotic mapping method for repeated folding and stretching transformation in a limited area. It is widely applied to pixel scrambling of images. The mathematical equation of Arnold mapping is given as follows:

$$\begin{bmatrix} X_{n+1} \\ Y_{n+1} \end{bmatrix} = \begin{bmatrix} 1 & a \\ b & ab+1 \end{bmatrix} \begin{bmatrix} X_n \\ Y_n \end{bmatrix} (\mathrm{mod}\,M) \tag{7}$$

where (X_n, Y_n) and (X_{n+1}, Y_{n+1}) represent the pixel coordinates of the original image and the pixel coordinates of transformed image, respectively. In addition, $X_n, Y_n, X_{n+1}, Y_{n+1} \in \{0, 1, \cdots, N-1\}$, and $a = 1, b = 1$. The variable M is the size of the image.

3.2. Security Analysis of Image Encryption Scheme

In this section, the Lena, fruits, and airplane gray images with a size of 256 $\times$ 256 are encrypted using the above image encryption scheme based on self-synchronous chaotic stream cipher. The security analysis results of image encryption are shown below.

3.2.1. Histogram Analysis

The histogram of the image is an important statistical feature of the image. It can be considered as the gray density function of the image. One of the evaluation criteria of image encryption effect is whether the gray histogram of the ciphertext image has the characteristics of uniform distribution. The results of the grayscale histogram are shown in Figure 5, and the horizontal and vertical coordinates of the gray histogram represent the pixel values and number of pixel values, respectively. As can be seen from the Figure 5g–i, the pixel values of the ciphertext images are so evenly distributed that it is difficult for the attacker to extract the plaintext pixel statistical characteristics from the ciphertext. Therefore, this image encryption scheme can resist statistical attacks well.

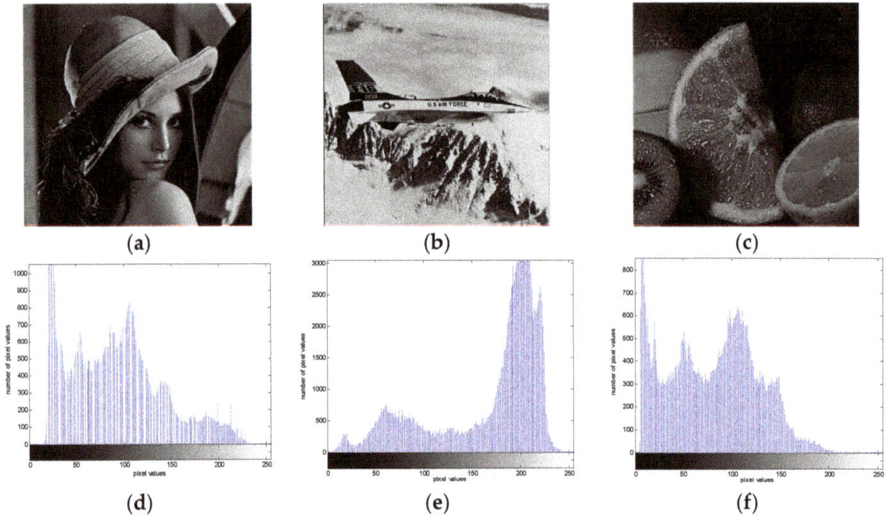

Figure 5. *Cont.*

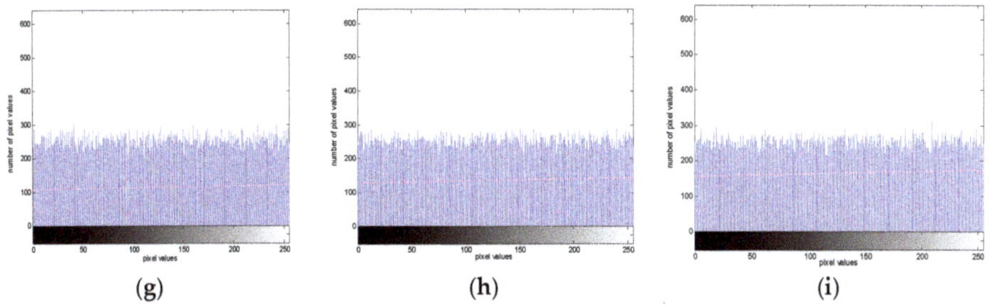

Figure 5. Histogram test with: (**a**) plain-image of Lena; (**b**) plain-image of airplane; (**c**) plain-image of fruits; (**d**) histogram of plain-image of Lena; (**e**) histogram of plain-image of airplane; (**f**) histogram of plain-image of fruits; (**g**) histogram of encrypted image of Lena; (**h**) histogram of encrypted image of airplane; (**i**) histogram of encrypted image of fruits.

3.2.2. Correlation Analysis of Adjacent Pixels

Correlation analysis refers to the analysis of two variables with correlation so as to measure the correlation degree of two variables. The correlation of adjacent pixels can reflect the scrambling effect of image pixels. The mathematical equation for the correlation of adjacent pixels is shown as follows [26]:

$$E(x) = \frac{1}{N} \sum_{k=1}^{N} x_k \tag{8}$$

$$D(x) = \frac{1}{N} \sum_{k=1}^{N} (x_k - E(x))^2 \tag{9}$$

$$\mathrm{cov}(x,y) = \frac{1}{N} \sum_{k=1}^{N} (x_k - E(x))(y_k - E(y)) \tag{10}$$

$$\rho_{xy} = \frac{\mathrm{cov}(x,y)}{\sqrt{D(x)}\sqrt{D(y)}} \tag{11}$$

where x_k and y_k represent the grey values of two adjacent pixels, and N is the number of randomly selected adjacent pixels from the original or encrypted image. The ρ_{xy}, $E(x)$, $D(x)$ and $\mathrm{cov}(x,y)$ represent the correlation coefficient, mean value, variance, and covariance, respectively.

For the correlation analysis experiment, we randomly selected 2000 pairs of adjacent pixels in horizontal, vertical, and diagonal directions from the plain and encrypted images of Lena. The experimental results are shown in Figure 6. Where (i,j) represents the position coordinates of this pixel in the image. As can be seen from the figure, the correlation of cipher-images is much lower than that of plain-images. Furthermore, Table 3 shows the correlation analysis of the adjacent pixels with the Lena, airplane, and fruits image. Obviously, the correlation coefficient of the plain-image is close to 1. On the contrary, the correlation coefficient of the cipher-image is close to 0, which indicates a good performance.

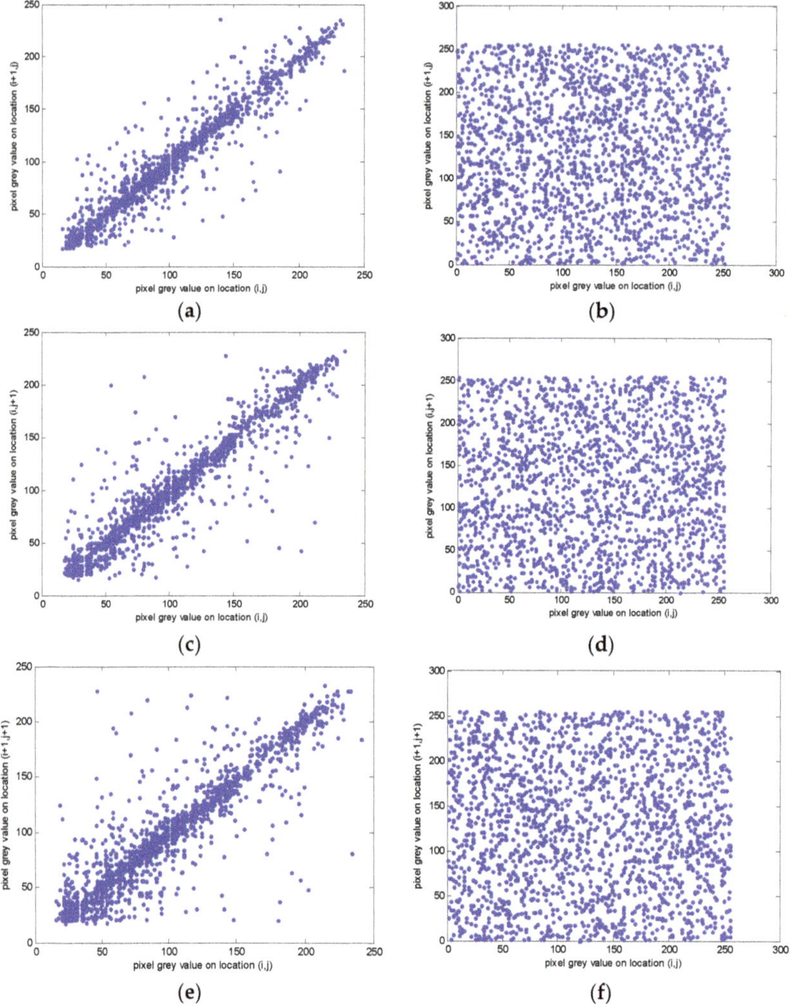

Figure 6. The correlation plots of two adjacent pixels for the plain and encrypted images of Lena with: (**a**) horizontal correlation of plain-image of Lena; (**b**) horizontal correlation of cipher-image of Lena; (**c**) vertical correlation of plain-image of Lena; (**d**) vertical correlation of cipher-image of Lena; (**e**) diagonal correlation of plain-image of Lena; (**f**) diagonal correlation of cipher-image of Lena.

Table 3. Correlation analysis of adjacent pixels for the Lena, airplane, and fruits images.

Direction	Horizontal	Vertical	Diagonal
Plain-image of Lena	0.9577	0.9440	0.9126
Cipher-image of Lena	−0.0082	0.0027	0.0030
Plain-image of Airplane	0.9147	0.9225	0.9109
Cipher-image of Airplane	0.0334	−0.0285	−0.0073
Plain-image of Fruits	0.9540	0.9497	0.9459
Cipher-image of Fruits	−0.0273	−0.0176	−0.0026

3.2.3. Peak Signal-To-Noise Ratio (PSNR) Analysis

The peak signal-to-noise ratio is an objective criterion for evaluating images, and its mathematical equation is given as follows:

$$\text{PSNR} = 10 \log_{10}(L^2/\text{MSE}) \tag{12}$$

$$\text{MSE} = \frac{1}{M^2} \sum_{i=1}^{M} \sum_{j=1}^{M} \left(I'(i,j) - I(i,j) \right)^2 \tag{13}$$

where $M \times M$ is the size of image, $I'(i,j)$, and $I(i,j)$ represent the pixel value of encrypted and original images, respectively. MSE is mean squared error, and L is the range of gray values in the image. Generally speaking, the better the encryption effect is, the smaller the PSNR of the image becomes. The results of the PSNR test is shown in Table 4. For the test result of our scheme, it is the average PSNR of the Lena, airplane, and fruits images. Similarly, for the image encryption method of Yin et al. [27] and Zhu [28], this PSNR represents the average value of multiple images. As can be seen from the table, our scheme has a smaller PSNR value, which shows a good encryption effect.

Table 4. Peak signal-to-noise ratio (PSNR) test with different methods.

Methods	PSNR Value
Our scheme	8.1543
Yin et al. [27]	8.4100
Zhu [28]	9.2322

3.2.4. Information Entropy Analysis

Information entropy can measure the distribution of gray values in images. The more random the gray value distribution, the greater the information entropy of the image. According to the information theory of Shannon, the information entropy can be defined as follows:

$$H(s) = -\sum_{i=0}^{2^n-1} P(s_i) \log_2(P(s_i)) \tag{14}$$

where $P(s_i)$ represents the probability of symbol s_i, n is the number of bits required to store each pixel value, 2^n is the total states of the information source s. When $n = 8$, the theoretical value of information entropy is 8. The information entropy test was performed for the encrypted images of Lena, airplane, and fruits. Table 5 shows the experimental results of the entropy test. The test result of our scheme is the average entropy value of the Lena, airplane, and fruits images. For the scheme of Liu et al. [29], Liu et al. [30], and Niyat [31], these entropy values represent the average value of multiple images. As can be seen from the table, the entropy value of encrypted image of our scheme is closer to the theoretical value 8. Therefore, this scheme can effectively resist the information entropy attack.

Table 5. Information entropy test with different methods.

Methods	Entropy Value
Our scheme	7.9971
Liu et al. [29]	7.9914
Liu et al. [30]	7.9851
Niyat et al. [31]	7.9877

4. Discussion

In this paper, we proposed a novel image encryption scheme with the purpose of ensuring secure transmission of image data. A four-dimensional hyperchaotic system is constructed to act as the chaotic sequence generator. Moreover, the performance of chaotic binary sequences is analyzed by multi-scale permutation entropy and the NIST-800-22 test. The test results show that the binary sequence has good randomicity and security. On the basis of the chaotic sequence generator, a self-synchronous stream cipher is designed to encrypt the image data. The generation of the key stream of the chaotic stream cipher is related to the seed key and a certain number of bits ciphertext that has been generated previously. It has the advantages of limited error propagation, self-synchronous and ciphertext statistical diffusion, which can satisfy the security and stability of image encryption system. Finally, a novel image encryption scheme is designed based on the self-synchronous stream cipher. Arnold mapping and wavelet transform are used to obtain a good scrambling effect for the digital image. Some simulation results show that the proposed scheme is both reliable and secure.

Author Contributions: C.F. conceived and wrote the paper, which devoted to the simulation experiments. Q.D. gave some theoretical guidance. All authors have read and approved the final manuscript.

Acknowledgments: This work was supported by Natural Science Foundation of China (No. 61471158) and "modern sensing technology" innovation team project of Heilongjiang province (No. 2012TD007).

Conflicts of Interest: The authors declare no conflict of interest.

References

1. Asikuzzaman, M.; Pickering, M.R. An overview of digital video watermarking. *IEEE Trans. Circuit Syst. Video Technol.* **2017**, *27*, 1–23. [CrossRef]
2. Kengne, J. Coexistence of chaos with hyperchaos, period-3 doubling bifurcation, and transient chaos in the hyperchaotic oscillator with gyrators. *Int. J. Bifurc. Chaos* **2015**, *25*, 1550052. [CrossRef]
3. Mahmoud, E.E.; Abood, F.S. A new nonlinear chaotic complex model and its complex antilag synchronization. *Complexity* **2017**, *25*, 3848953. [CrossRef]
4. Wang, Q.X.; Yu, S.M.; Li, C.Q.; Lu, J.H.; Fang, X.L.; Guyeux, C.; Bahi, J.M. Theoretical design and FPGA-based implementation of higher-dimensional digital chaotic systems. *IEEE Trans. Circuit Syst. I* **2016**, *63*, 401–412. [CrossRef]
5. Yang, Q.G.; Chen, G.R. A chaotic system with one saddle and two stable node-foci. *Int. J. Bifurc. Chaos* **2008**, *18*, 1393–1414. [CrossRef]
6. Li, C.H.; Luo, G.C.; Qin, K.; Li, C.B. An image encryption scheme based on chaotic tent map. *Nonlinear Dyn.* **2017**, *87*, 127–133. [CrossRef]
7. Cao, L.C.; Luo, Y.L.; Qiu, S.H.; Liu, J.X. A perturbation method to the tent map based on Lyapunov exponent and its application. *Chin. Phys. B* **2015**, *24*, 326–337. [CrossRef]
8. Usama, M.; Khan, M.K.; Alghathbar, K.; Lee, C. Chaos-based secure satellite imagery cryptosystem. *Comput. Math. Appl.* **2010**, *60*, 326–337. [CrossRef]
9. Zheng, Y.F.; Jin, J.X. A novel image encryption scheme based on Henon map and compound spatiotemporal chaos. *Multimed. Tools Appl.* **2015**, *74*, 7803–7820. [CrossRef]
10. Ping, P.; Xu, F.; Mao, Y.C.; Wang, Z.J. Designing permutation-substitution image encryption networks with Henon map. *Neurocomputing* **2018**, *283*, 53–63. [CrossRef]
11. Ye, G.D.; Huang, X.L. An efficient symmetric image encryption algorithm based on an intertwining logistic map. *Neurocomputing* **2017**, *251*, 45–53. [CrossRef]
12. Haroun, M.F.; Gulliver, T.A. Real-time image encryption using a low-complexity discrete 3D dual chaotic cipher. *Nonlinear Dyn.* **2015**, *82*, 1523–1535. [CrossRef]
13. Persohn, K.J.; Povinelli, R.J. Analyzing logistic map pseudorandom number generators for periodicity induced by finite precision floating-point representation. *Chaos Soliton Fract.* **2012**, *45*, 238–245. [CrossRef]
14. Zheng, Y.B.; Song, Y.; Du, B.X.; Pan, J.; Ding, Q. A novel detection of periodic phenomena of binary chaotic sequences. *Acta. Phys. Sin. Chin. Ed.* **2012**, *61*, 230501.
15. Khelifi, F. On the security of a stream cipher in reversible data hiding schemes operating in the encrypted domain. *Signal Process.* **2018**, *143*, 336–345. [CrossRef]
16. Ullagaddi, V.; Hassan, F.; Devabhaktuni, V. Symmetric synchronous stream encryption using images. *Signal Image Video Process.* **2015**, *9*, 1–8. [CrossRef]

17. Bucerzan, D.; Craciun, M.; Chis, V.; Ratiu, C. Stream ciphers analysis methods. *Int. J. Comput. Commun.* **2010**, *5*, 483–489. [CrossRef]

18. Chen, G.R.; Lai, D.J. Making a dynamical system chaotic: Feedback control of Lyapunov exponents for discrete-time dynamical systems. *IEEE Trans. Circuit Syst. I Fundam. Theory Appl.* **2002**, *44*, 250–253. [CrossRef]

19. Chen, G.R.; Lai, D.J. Feedback control of lyapunov exponents for discrete-time dynamical systems. *Int. J. Bifurc. Chaos* **1996**, *6*, 1341–1349. [CrossRef]

20. Zhang, X.Y.; Liang, Y.T.; Zhou, J.Z.; Zang, Y. A novel bearing fault diagnosis model integrated permutation entropy, ensemble empirical mode decomposition and optimized SVM. *Measurement* **2015**, *69*, 164–179. [CrossRef]

21. Bandt, C.; Pompe, B. Permutation entropy: A natural complexity measure for time series. *Phys. Rev. Lett.* **2002**, *88*, 174102. [CrossRef] [PubMed]

22. Sun, K.H.; He, S.B.; Yin, L.Z.; Duo, L.K. Application of FuzzyEn algorithm to the analysis of complexity of chaotic sequence. *Acta Phys. Sin. Chin. Ed.* **2012**, *61*, 130507.

23. Xu, W.; Ding, Q.; Zhang, X.G. Improving the complexity of chaotic sequence based on the PCA algorithm. *J. Appl. Anal. Comput.* **2015**, *5*, 262–272.

24. Rukhin, A.; Soto, J.; Nechvatal, J.; Miles, S.; Barker, E. A statistical test suite for random and pseudorandom number generators for cryptographic applications. *Appl. Phys. Lett.* **2015**, *22*, 1645–1776.

25. Soleymani, A.; Nordin, M.J.; Sundararajan, E. A chaotic cryptosystem for images based on Henon and Arnold cat map. *Sci. World J.* **2014**, *2014*, 536930. [CrossRef] [PubMed]

26. Li, T.Y.; Yang, M.G.; Wu, J.; Jing, X. A novel image encryption algorithm based on a fractional-order hyperchaotic system and DNA computing. *Complexity* **2017**, *2017*, 9010251. [CrossRef]

27. Yin, Q.; Wang, C.H. A new chaotic image encryption scheme using breadth-first search and dynamic diffusion. *Int. J. Bifurc. Chaos* **2018**, *28*, 1850047. [CrossRef]

28. Zhu, C.X. A novel image encryption scheme based on improved hyperchaotic sequences. *Opt. Commun.* **2012**, *285*, 29–37. [CrossRef]

29. Liu, H.J.; Kadir, A.; Sun, X.B. Chaos-based fast colour image encryption scheme with true random number keys from environmental noise. *IET Image Process.* **2017**, *11*, 324–332. [CrossRef]

30. Liu, H.J.; Wang, X.Y. Color image encryption based on one-time keys and robust chaotic maps. *Comput. Math. Appl.* **2010**, *59*, 3320–3327. [CrossRef]

31. Niyat, A.Y.; Moattar, M.H.; Torshiz, M.N. Color image encryption based on hybrid hyper-chaotic system and cellular automata. *Opt. Laser Eng.* **2017**, *90*, 225–237. [CrossRef]

Article

The General Solution of Singular Fractional-Order Linear Time-Invariant Continuous Systems with Regular Pencils

Iqbal M. Batiha [1], Reyad El-Khazali [2,], Ahmed AlSaedi [3] and Shaher Momani [1]

[1]   Department of Mathematics, Faculty of Science, University of Jordan, Amman 11942, Jordan; aqd9140074@ju.edu.jo (I.M.B.); s.momani@ju.edu.jo (S.M.)
[2]   ECCE Department, Khalifa University, Abu-Dhabi 127788, United Arab Emirates
[3]   Department of Nonlinear Analysis and Applied Mathematics (NAAM) Research Group, Faculty of Science, King Abdulaziz University (KAU), Jeddah 21589, Saudi Arabia; aalsaedi@kau.edu.sa
*   Correspondence: reyad.elkhazali@ku.ac.ae or reyad.elkhazali@kustar.ac.ae

Received: 18 April 2018; Accepted: 19 May 2018; Published: 23 May 2018

Abstract: This paper introduces a general solution of singular fractional-order linear-time invariant (FoLTI) continuous systems using the Adomian Decomposition Method (ADM) based on the Caputo's definition of the fractional-order derivative. The complexity of their entropy lies in defining the complete solution of such systems, which depends on introducing a method of decomposing their dynamic states from their static states. The solution is formulated by converting the singular system of regular pencils into a recursive form using the sequence of transformations, which separates the dynamic variables from the algebraic variables. The main idea of this work is demonstrated via numerical examples.

Keywords: fractional calculus; Adomian decomposition; Mittag–Leffler function; descriptor fractional linear systems; regular pencils; Schur factorization

---

1. Introduction

A dynamical system represented by differential equations with non-integer order derivatives is denoted as a fractional-order system. In general, most practical systems are best described by fractional-order dynamics (FoD), where the integer-order representation of such systems is considered as a special case. Recently, different types of problems of fractional-order dynamical systems have been considered in the literature [1,2]. Time-domain system identification using the fractional-order models was initiated in the late nineties. Several methods of discretizing the fractional-order differential equation using Grunwald–Letnikov (GL) approximation or phase assignment technique can be found in [3,4], while another biquadratic approximation of the fractional-order Laplacian operator based on the flatness of the phase frequency response at its center frequency is discussed in [5]. Furthermore, the state–space representations of fractional-order systems have been broadly used to investigate system stability, observability and controllability [6–8]. The generalization of FoD have allowed it to flourish in many fields of applications, such as control theory, communication systems and applied mathematics [9,10].

The singular (descriptor) fractional-order system of differential equations plays an important role in many applications, such as electric networks, economics, optimization problems, analysis of control systems, constrained mechanics, aircraft and robot dynamics, biology and large-scale systems [9,10]. Many continuous or discrete-time systems are usually described by complete dynamical states that vary with time, which have wide applications in social sciences, chaotic systems, economics, electrical networks, information theory and medical sciences [11–16]. Since the singular systems enjoy static and dynamic states, the complexity of their entropy depends on the methods of decomposing these states from each other to completely identify the analytical solution of such systems.

The solution of singular systems with regular and singular pencils was discussed in references [17–21], while the optimal solution of a class of singular linear systems of regular and singular pencils that have non-consistent linear systems of nabla difference equations with non-consistent initial conditions was discussed in reference [22]. The relationship between the solutions of an initial value problem of a linear singular system of fractional nabla difference equations, its proper dual system and its transposed dual system as well as introduced necessary and sufficient conditions for the existence and uniqueness of their solution were thoroughly investigated in reference [23]. The initial value problem of a class of non-homogeneous singular systems of fractional nabla difference equations with constant matrix coefficients was investigated in reference [24], which considered two cases: square coefficient matrices with a singular leading coefficient and regular pencils; and square and non-square matrices of singular pencils.

In this work, we only considered singular linear systems with regular pencils. The case of the systems of singular pencils is left for further development. To find the general solution of fractional-order singular systems, the Adomian Decomposition Method (ADM) [22,23] is extended by first introducing the general solution of the regular commensurate fractional-order linear-time invariant (FoLTI) continuous systems, which is described by the following general form:

$$D^\alpha x(t) = Ax(t) + Bu(t), 0 < \alpha \leq 1, \tag{1a}$$

$$y(t) = Cx(t) + Du(t), \tag{1b}$$

where $x(t) \in \mathbb{R}^n$, $u(t) \in \mathbb{R}^m$ and $y(t) \in \mathbb{R}^p$ are the system states, the input, and output vectors, respectively; while $A \in \mathbb{R}^{n \times n}$, $B \in \mathbb{R}^{n \times m}$, $C \in \mathbb{R}^{p \times n}$ and $D \in \mathbb{R}^{p \times m}$ are the system constant matrices.

The ADM is extended here to obtain the solution of a singular FoLTI continuous system that has the following general form:

$$ED^\alpha x(t) = Ax(t) + Bu(t), 0 < \alpha \leq 1, \tag{2a}$$

$$y(t) = Cx(t) + Du(t), \tag{2b}$$

where $E \in \mathbb{R}^{n \times n}$ is a singular matrix; $x(t) \in \mathbb{R}^n$ is the pseudo-state; $u(t) \in \mathbb{R}^m$ is the control input; $y(t) \in \mathbb{R}^p$ is the output; and $A \in \mathbb{R}^{n \times n}$, $B \in \mathbb{R}^{n \times m}$ and $C \in \mathbb{R}^{p \times n}$ with rank $C = p$.

Definition 1. [24] *The matrix pencil $sE - A$, where E and $A \in \mathbb{R}^{n \times q}$ and for an arbitrary $s \in \mathbb{C}$, is called:*

(1) *Regular when $n = q$ and $\det (sE - A) \neq 0$,*
(2) *Singular when $n = q$ or $n \neq q$ and $\det (sE - A) = 0$.*

In this work, we considered the class of regular pencils with a singular matrix E. Singular fractional-order systems consist of coupled differential and algebraic equations. The control of singular fractional-order systems is not well-flourished compared to that of the conventional dynamical systems. However, it is possible to use the sequence of transformations to decouple the differential and the algebraic parts of the system from each other, thus enabling the application of the standard state–space control theory to a dynamical subsystem of a lower order [24–27].

There are three main steps used to decouple the system's static and dynamic parts from each other. The first one involves using the generalized Schur decomposition method, the second one involves solving a coupled Sylvester equation and the third one involves constructing well-defined transformation matrices [28,29]. The first step is thoroughly investigated using numerical linear algebra. Various existing methods for transforming a matrix into a Jordan-Schur form and a matrix pencil into a Weierstrass–Schur form have been investigated in reference [28]. These methods are extended to extract the partial information that corresponds to the dominant eigenvalues from large-scale matrices and matrix pencils. The solution and perturbation analysis of a coupled Sylvester equation is presented in reference [29]. The Schur method and the Hessenberg–Schur method are extended for a coupled

Sylvester equation, which is transformed into a standard Sylvester equation. This equation is solved using standard techniques presented in [28,29].

This work is outlined as follows. In the next section, the necessary definitions and preliminaries are introduced. Section 3 describes the ADM method. Section 4 introduces the solution of FoLTI systems with regular pencils. A recursive method to decompose the singular systems is introduced in Section 5, followed by numerical examples in Section 6. The summaries and concluding remarks are presented in Section 7.

2. Basic Definitions and Preliminaries

The Caputo definition of fractional-order derivatives is adopted in this work. It is a modification of the Riemann–Liouville definition and it has the advantage of only using the initial conditions that corresponds to integer-order derivatives, which is suitable for most physical systems [30–32]. The following definitions and preliminaries of fractional-order calculus are presented here for completeness.

Definition 2 [9,10]. *Let $f(t)$ be an integrable piecewise continuous function on any finite subinterval of $(0, +\infty)$. Thus, the fractional integral of $f(t)$ of order α is defined as:*

$$J^{\alpha} f(t) := \frac{t^{\alpha-1}}{\Gamma(\alpha)} \times f(t) = \frac{1}{\Gamma(\alpha)} \int_0^t (t - \tau)^{\alpha-1} f(\tau) d\tau, \ t > 0, \ \alpha > 0. \tag{3}$$

In this paper, we will use the following equality [20]:

$$J^{\alpha} t^{\mu} = \frac{\Gamma(\mu + 1)}{\Gamma(\mu + \alpha + 1)} t^{\mu+\alpha}, \ \alpha > 0, \ \mu > -1, t > 0. \tag{4}$$

Definition 3 [9,10]. *The Caputo fractional-order derivative is defined as:*

$$D^{\alpha} f(t) = \frac{1}{\Gamma(M - \alpha)} \int_0^t \frac{f^M(\tau)}{(t - \tau)^{\alpha+1-M}} d\tau, f^M(\tau) = \frac{d^M f(\tau)}{d\tau^M}, \tag{5}$$

where $\Gamma(\cdot)$ is the Gamma function and $M - 1 \leq \alpha < M, \ M \in \mathbb{N}$.

Definition 4 [8]. *The Mittag–Leffler function of two parameters is defined by:*

$$E_{\alpha,\beta}(t) = \sum_{k=0}^{\infty} \frac{t^k}{\Gamma(k\alpha + \beta)}. \tag{6}$$

Definition 5 [9]. *The Mittag–Leffler matrix function of two parameters is defined by:*

$$E_{\alpha,\beta}(At^{\alpha}) = \sum_{k=0}^{\infty} \frac{A^k t^{\alpha k}}{\Gamma(k\alpha + \beta)}, \tag{7}$$

where $A \in \mathbb{R}^{n \times n}$.

Definition 6 [31]. *The system given by (1) or the pair (E, A) is said to be regular pencil if there exists a unique solution $x(t)$ for a given initial condition.*

Lemma 1 [32]. *The system described by (1) or the pair (E, A) is said to be a regular pencil if and only if $det(Es^{\alpha} - A) \neq 0$, for $s \in \mathbb{C}$.*

3. Solution of FoLTI Systems Using ADM Method

In this section, we used the ADM method to obtain the general solution of fractional-order state equations of linear time-invariant continuous systems. See [25–27] for an overview of the ADM technique. In the subsequent discussion, consider the linear system described by (1a) and assume the definition of Caputo fractional-order derivative; i.e.,

$$D^\alpha x(t) = Ax(t) + Bu(t), 0 < \alpha \le 1, \tag{8}$$

with the initial condition:

$$x(0) = v. \tag{9}$$

Notice that applying J^α (i.e., fractional-order integration of order α) on both sides of system (8) yields:

$$x(t) = x(0) + AJ^\alpha x(t) + BJ^\alpha u(t). \tag{10}$$

To use the Adomian decomposition method, we assume that the general solution of (8) takes the general form of $x(t) = \sum_{k=0}^\infty x_k(t)$, in which:

$$x_0(t) = v + BJ^\alpha u(t) \tag{11}$$

and

$$x_k(t) = AJ^\alpha x_{k-1}(t), \ k \ge 1. \tag{12}$$

Now, from (11) and (12), one can obtain the following recursive formula for the system states:

$$x_1(t) = J^\alpha[Av + ABJ^\alpha u(t)], \ x_2(t) = J^{2\alpha}[A^2v + A^2BJ^\alpha u(t)], \ \dots, \ x_k(t) = J^{k\alpha}\left[A^k v + A^k BJ^\alpha u(t)\right]. \tag{13}$$

Therefore,

$$x_1(t) = \frac{Av}{\Gamma(\alpha+1)}t^\alpha + ABJ^{2\alpha}u(t), \ x_2(t) = \frac{A^2v}{\Gamma(2\alpha+1)}t^{2\alpha} + A^2BJ^{3\alpha}u(t), \ \dots,$$

$$\tag{14}$$

$$x_k(t) = \frac{A^k v}{\Gamma(k\alpha+1)}t^{k\alpha} + A^k BJ^{(k+1)\alpha}u(t).$$

Since the general solution $x(t) = \sum_{k=0}^\infty x_k(t)$, Equation (14) then yields:

$$x(t) = \sum_{k=0}^\infty \frac{(At^\alpha)^k}{\Gamma(k\alpha+1)}v + \sum_{k=0}^\infty A^k BJ^{(k+1)\alpha}u(t). \tag{15}$$

That is,

$$x(t) = \sum_{k=0}^\infty \frac{(At^\alpha)^k}{\Gamma(k\alpha+1)}v + \sum_{k=0}^\infty A^k B\frac{1}{\Gamma((k+1)\alpha)} \int_0^t (t-\tau)^{(k+1)\alpha-1}u(\tau)d\tau. \tag{16}$$

or,

$$x(t) = \sum_{k=0}^\infty \frac{(At^\alpha)^k}{\Gamma(k\alpha+1)}v + \int_0^t \sum_{k=0}^\infty \frac{A^k(t-\tau)^{(k+1)\alpha-1}}{\Gamma((k+1)\alpha)}Bu(\tau)d\tau. \tag{17}$$

Alternately, in terms of the Mittag–Leffler matrix functions of (7), one may rewrite (17) as follows:

$$x(t) = E_{\alpha,1}(At^\alpha)v + \left[t^{\alpha-1}E_{\alpha,\alpha}(At^\alpha)\right][Bu(\tau)]. \tag{18}$$

Consequently, the general solution of system states, described by (8), can be written in the following general form:

$$x(t) = \phi_{\alpha 0}(t)v + \int_0^t \phi_\alpha(t-\tau)Bu(\tau)d\tau, x(0) = v \tag{19}$$

where

$$\phi_{\alpha 0}(t) = E_{\alpha,1}(At^{\alpha}) = \sum_{k=0}^{\infty} \frac{(At^{\alpha})^k}{\Gamma(k\alpha + 1)}, \tag{20}$$

$$\phi_{\alpha}(t) = t^{\alpha-1} E_{\alpha,\alpha}(At^{\alpha}) = \sum_{k=0}^{\infty} \frac{A^k t^{(k+1)\alpha-1}}{\Gamma((k+1)\alpha)}. \tag{21}$$

and the general solution of the system output is given by:

$$y(t) = C\left\{ E_{\alpha,1}(At^{\alpha})v + \left[t^{\alpha-1} E_{\alpha,\alpha}(At^{\alpha}) \right] [Bu(t)] \right\} + Du(t). \tag{22}$$

4. The General Solution of FoLTI Singular Systems with Regular Pencils

The general solution of FoLTI singular continuous systems is usually obtained by first transforming the system into the canonical form [33–35], which enables one to easily decompose the static terms from the dynamic ones. The following lemmas are presented for completeness to derive the general solution of the FoLTI singular systems with regular pencils. To simplify the process of obtaining the general solution, the system matrices (E, A) with regular pencils may be both transformed into a triangular form with the zero eigenvalues of E placed at the lower right block.

Lemma 2 [33]. *Consider the following FoLTI singular continuous system:*

$$ED^{\alpha}x(t) = Ax(t) + Bu(t), 0 < \alpha \leq 1, \tag{23}$$

where $E \in \mathbb{R}^{n\times n}$ is a singular matrix of rank $n_1 < n$, $x(t) \in \mathbb{R}^n$, $u(t) \in \mathbb{R}^m$, and $A \in \mathbb{R}^{n\times n}$, $B \in \mathbb{R}^{n\times m}$..

If (23) is regular, there exist non-singular matrices P_1, $Q_1 \in \mathbb{R}^{n\times n}$, such that:

$$P_1 E Q_1 = \begin{bmatrix} E_1 & E_2 \\ 0 & E_3 \end{bmatrix}, P_1 A Q_1 = \begin{bmatrix} J_1 & J_2 \\ 0 & J_3 \end{bmatrix}, \tag{24}$$

where $E_1 \in \mathbb{R}^{n_1 \times n_1}$ is non-singular; $E_3 \in \mathbb{R}^{n_2 \times n_2}$ is an upper triangular matrix with all diagonal elements being zero; $J_1 \in \mathbb{R}^{n_1 \times n_1}$; $J_3 \in \mathbb{R}^{n_2 \times n_2}$ is non-singular and upper triangular; and E_2, $J_2 \in \mathbb{R}^{n_1 \times n_2}$.

The generalized Schur decomposition given by (24) and the subsequent reordering of the diagonal elements of E may be conducted using "*qz*" MATLAB function to construct E_1 and J_1 as upper triangular matrices [32].

Lemma 3 [36]. *Consider (24), then there exist matrices L, $R \in \mathbb{R}^{n_1 \times n_2}$, such that:*

$$\begin{bmatrix} I & L \\ 0 & I \end{bmatrix}\begin{bmatrix} E_1 & E_2 \\ 0 & E_3 \end{bmatrix}\begin{bmatrix} I & R \\ 0 & I \end{bmatrix} = \begin{bmatrix} E_1 & 0 \\ 0 & E_3 \end{bmatrix} \tag{25}$$

and

$$\begin{bmatrix} I & L \\ 0 & I \end{bmatrix}\begin{bmatrix} J_1 & J_2 \\ 0 & J_3 \end{bmatrix}\begin{bmatrix} I & R \\ 0 & I \end{bmatrix} = \begin{bmatrix} J_1 & 0 \\ 0 & J_3 \end{bmatrix}. \tag{26}$$

Lemma 4 [36]. *Consider system (23), if this system is regular, then there exist non-singular matrices P, $Q \in \mathbb{R}^{n\times n}$ such that the transformation:*

$$PEQQ^{-1}D^{\alpha}x(t) = PAQQ^{-1}x(t) + PBu(t) \tag{27}$$

yields the following structure:

$$\begin{bmatrix} I_{n_1} & 0 \\ 0 & N \end{bmatrix} Q^{-1} D^\alpha x(t) = \begin{bmatrix} F & 0 \\ 0 & I_{n_2} \end{bmatrix} Q^{-1} x(t) + \begin{bmatrix} H \\ K \end{bmatrix} u(t), \tag{28}$$

where $N \in \mathbb{R}^{n_2 \times n_2}$ is a nilpotent matrix, $F \in \mathbb{R}^{n_1 \times n_1}$; and n_1 is equal to the degree of the polynomial $\det(Es - A)$, such that $n_1 + n_2 = n$, and where:

$$PEQ = \begin{bmatrix} I_{n_1} & 0 \\ 0 & N \end{bmatrix}, \quad PAQ = \begin{bmatrix} F & 0 \\ 0 & I_{n_2} \end{bmatrix}, \quad PB = \begin{bmatrix} H \\ K \end{bmatrix}, \tag{29}$$

where $H \in \mathbb{R}^{n_1 \times m}$, $K \in \mathbb{R}^{n_2 \times m}$ and $s \in \mathbb{C}$.

Proof. According to Lemma 1, let P_1 and Q_1 be matrices such that $P = P_3 P_2 P_1$ and $Q = Q_1 Q_2$, where:

$$P_2 \equiv \begin{bmatrix} I & L \\ 0 & I \end{bmatrix}, \quad P_3 \equiv \begin{bmatrix} E_1^{-1} & R \\ 0 & J_3^{-1} \end{bmatrix}, \quad \text{and } Q_2 \equiv \begin{bmatrix} I & R \\ 0 & I \end{bmatrix}, \tag{30}$$

where $P_2, P_3, Q_2 \in \mathbb{R}^{n \times n}$. $\square$

Now from Lemma 2, the matrices L and R satisfy:

$$PEQ = \begin{bmatrix} I & L \\ 0 & J_3^{-1} E_3 \end{bmatrix}, \quad PAQ = \begin{bmatrix} E_1^{-1} J_1 & R \\ 0 & I \end{bmatrix}, \tag{31}$$

where $N = J_3^{-1} E_3$ is a nilpotent matrix because E_3 is an upper triangular matrix with zero diagonal elements; while J_3^{-1} and J_3 are both upper triangular matrices. The form of (28) is obtained by letting $F = E_1^{-1} J_1$.

Now consider system (2) with $D = 0$, i.e.,

$$ED^\alpha x(t) = Ax(t) + Bu(t), 0 < \alpha \leq 1, x(0) = v \tag{32}$$

$$y(t) = Cx(t) \tag{33}$$

Let the system of (32) be regular. From Lemma 4, it follows that (32) can be rewritten as [37,38]:

$$\begin{bmatrix} I_{n_1} & 0 \\ 0 & N \end{bmatrix} \begin{bmatrix} D^\alpha w_1(t) \\ D^\alpha w_2(t) \end{bmatrix} = \begin{bmatrix} F & 0 \\ 0 & I_{n_2} \end{bmatrix} \begin{bmatrix} w_1(t) \\ w_2(t) \end{bmatrix} + \begin{bmatrix} H \\ K \end{bmatrix} u(t), \tag{34}$$

$$w(t) = Q^{-1} x(t) \equiv \begin{bmatrix} w_1(t) & w_2(t) \end{bmatrix}^T, \tag{35}$$

where $w_1(t) \in \mathbb{R}^{n_1}$; $w_2(t) \in \mathbb{R}^{n_2}$; and $PB = \begin{bmatrix} H & K \end{bmatrix}^T$.

Thus, the following two subsystems are obtained:

$$D^\alpha w_1(t) = F w_1(t) + H u(t), \tag{36}$$

$$N D^\alpha w_2(t) = w_2(t) + K u(t). \tag{37}$$

Since $x(0) = v$, therefore:

$$w_0 = Q^{-1} x(0) = Q^{-1} v \equiv [v_0 v_1]^T, \tag{38}$$

where $v_0(t) \in \mathbb{R}^{n_1}$ and $v_1(t) \in \mathbb{R}^{n_2}$.

The general solution of (36) is defined in terms of its fractional-order state equation (see (19)) as follows:

$$w_1(t) = \phi_{\alpha 0}(t)v + \int_0^t \phi_\alpha(t-\tau)Hu(\tau)d\tau, \; x(0) = v, \tag{39}$$

where

$$\phi_{\alpha 0}(t) = E_{\alpha,1}(Ft^\alpha) = \sum_{k=0}^\infty \frac{(Ft^\alpha)^k}{\Gamma(k\alpha+1)}, \tag{40a}$$

$$\phi_\alpha(t) = t^{\alpha-1}E_{\alpha,\alpha}(Ft^\alpha) = \sum_{k=0}^\infty \frac{F^k t^{(k+1)\alpha-1}}{\Gamma((k+1)\alpha)}. \tag{40b}$$

The solution of subsystem (37) is obtained using the property of the nilpotent matrix N. Thus, there are two cases to consider:

Case 1: $N = 0$,

$$\text{In this case, } w_2(t) = -Ku(t). \tag{41}$$

Case 2: $N \neq 0$,

To clarify this general case, let $N^2 = 0$. Pre-multiplying the second row of (34) by N yields:

$$N^2 D^\alpha w_2(t) = Nw_2(t) + NKu(t). \tag{42}$$

Now, differentiating both sides of (42) of order α (i.e., applying D^α to both sides) and using (35) yields:

$$w_2(t) = -Ku(t) - NKD^\alpha u(t) + N^2 D^{2\alpha} w_2(t). \tag{43}$$

Since $N^2 = 0$ by hypothesis, therefore:

$$w_2(t) = -Ku(t) - NKD^\alpha u(t). \tag{44}$$

In general, since N is a nilpotent matrix, there exists an integer number l such that $N^l = 0$. Now, pre-multiplying (34) by N^{l-1} and using (35) yields:

$$w_2(t) = -Ku(t) - \sum_{j=0}^{l-1} N^j KD^{j\alpha}u(t). \tag{45}$$

Substituting (40a) and (45) into (37) and (38) implies the following general solution of (34):

$$x(t) = Q\begin{bmatrix} I_{n_1} \\ 0_{n_2 \times n_1} \end{bmatrix}(\phi_{\alpha 0}(t)v_0 + \int_0^t \phi_\alpha(t-\tau)Hu(\tau)d\tau) + Q\begin{bmatrix} 0_{n_1 \times n_2} \\ I_{n_2} \end{bmatrix}\left(-Ku(t) - \sum_{j=0}^{l-1} N^j KD^{j\alpha}u(t)\right), \tag{46a}$$

where

$$\phi_{\alpha 0}(t) = E_{\alpha,1}(Ft^\alpha) = \sum_{k=0}^\infty \frac{(Ft^\alpha)^k}{\Gamma(k\alpha+1)}, \tag{46b}$$

$$\phi_\alpha(t) = t^{\alpha-1}E_{\alpha,\alpha}(At^\alpha) = \sum_{k=0}^\infty \frac{A^k t^{(k+1)\alpha-1}}{\Gamma((k+1)\alpha)}. \tag{46c}$$

Finally, the solution of $y(t)$ follows directly from (46a) and (34).

5. Recursive form of FoLTI Systems with Regular Pencils

A recursive solution of FoLTI systems with regular pencils is first investigated by considering the unforced system (homogeneous) of (32) and by using proper Schur transformation matrices. The following theorem outlines the main results of this work.

Theorem 1. *Consider a FoLTI homogeneous singular system with regular pencils of the following form:*

$$ED^\alpha x(t) = Ax(t), 0 < \alpha \leq 1, \tag{47a}$$

$$y(t) = Cx(t), \tag{47b}$$

where $E \in \mathbb{R}^{n \times n}$ is a singular matrix of rank $n - d < n$; $x(t) \in \mathbb{R}^n$; $y(t) \in \mathbb{R}^p$; $A \in \mathbb{R}^{n \times n}$; $C \in \mathbb{R}^{p \times n}$ with rank $C = p$; and with an initial condition $x(0) = v$. Let $A_\lambda \equiv (A - \lambda^\alpha E)^{-1} A$, $E_\lambda \equiv (A - \lambda^\alpha E)^{-1} E$; and let

$$Q = \begin{bmatrix} Q_s & \widetilde{Q_s} \end{bmatrix} \in \mathbb{R}^{n \times n} \text{ be a unitary matrix such that } E_\lambda = QTQ^* = \begin{bmatrix} Q_s & \widetilde{Q_s} \end{bmatrix} \begin{bmatrix} G_s & D_s \\ 0 & N_s \end{bmatrix} \begin{bmatrix} Q_s^* \\ \widetilde{Q_s}^* \end{bmatrix},$$

where $G_s \in \mathbb{R}^{n-d \times n-d}$ is an invertible matrix and $N_s \in \mathbb{R}^{d \times d}$ is a nilpotent matrix, which has all the zero eigenvalues of E_λ. Therefore, the general solution of (47) can be expressed as:

$$x(t) = Q_s E_{\alpha,1}\left(\left(G_s^{-1} + \lambda^\alpha I \right) t^\alpha \right) Q_s^* v = Q_s \left[\sum_{k=0}^\infty \frac{\left(\left(G_s^{-1} + \lambda^\alpha I \right) t^\alpha \right)^k}{\Gamma(k\alpha + 1)} \right] Q_s^* v \tag{48}$$

and

$$y(t) = Cx(t).$$

Proof. The general solution of (47) can be obtained by transforming it into a recursive form using the Schur factorization structure. Since the pair (E, A) is assumed to be a regular pencil, from Lemma 1, there exists some $\lambda \in \mathbb{C}$ such that $A - \lambda^\alpha E$ is invertible, which implies that $\det(A - \lambda^\alpha E) \neq 0$. Since $A_\lambda \equiv (A - \lambda^\alpha E)^{-1} A$, and $E_\lambda \equiv (A - \lambda^\alpha E)^{-1} E$ by hypothesis, pre-multiplying (47a) by $(A - \lambda^\alpha E)^{-1}$ yields:

$$(A - \lambda^\alpha E)^{-1} E D^\alpha x(t) = (A - \lambda^\alpha E)^{-1} A \, x(t). \tag{49}$$

System (47) can be rewritten as:

$$E_\lambda D^\alpha x(t) = A_\lambda x(t). \tag{50}$$

Since

$$A_\lambda = (A - \lambda^\alpha E)^{-1} A = (A - \lambda^\alpha E)^{-1}(A - \lambda^\alpha E + \lambda^\alpha E) = I + \lambda^\alpha E_\lambda, \tag{51}$$

Then (50) is defined as:

$$E_\lambda D^\alpha x(t) = (I + \lambda^\alpha E_\lambda) x(t). \tag{52}$$

Now, since $E_\lambda = QTQ^* = \begin{bmatrix} Q_s & \widetilde{Q_s} \end{bmatrix} \begin{bmatrix} G_s & D_s \\ 0 & N_s \end{bmatrix} \begin{bmatrix} Q_s^* \\ \widetilde{Q_s}^* \end{bmatrix}$ by hypothesis, which is verified in [35], one may decompose the system states of (47a) to obtain:

$$x(t) = Q_s x_1(t) + \widetilde{Q_s} x_2(t), \tag{53}$$

where $x_1 \in \mathbb{R}^{n-d}$, $x_2 \in \mathbb{R}^d$; $d > 0$.

Substituting (53) into (52) gives:

$$E_\lambda \begin{bmatrix} Q_s & \widetilde{Q_s} \end{bmatrix} D^\alpha \begin{bmatrix} x_1(t) \\ x_2(t) \end{bmatrix} = (I + \lambda^\alpha E_s) \begin{bmatrix} Q_s & \widetilde{Q_s} \end{bmatrix} \begin{bmatrix} x_1(t) \\ x_2(t) \end{bmatrix}. \tag{54}$$

Moreover, applying the Schur decomposition on (52) yields:

$$\begin{bmatrix} Q_s & \widetilde{Q_s} \end{bmatrix} \begin{bmatrix} G_s & D_s \\ 0 & N_s \end{bmatrix} \begin{bmatrix} Q_s^* \\ \widetilde{Q_s}^* \end{bmatrix} \begin{bmatrix} Q_s & \widetilde{Q_s} \end{bmatrix} \begin{bmatrix} D^\alpha x_1(t) \\ D^\alpha x_2(t) \end{bmatrix} = \begin{bmatrix} Q_s & \widetilde{Q_s} \end{bmatrix} \begin{bmatrix} x_1(t) \\ x_2(t) \end{bmatrix} + \begin{bmatrix} Q_s & \widetilde{Q_s} \end{bmatrix} \begin{bmatrix} \lambda^\alpha G_s & \lambda^\alpha D_s \\ 0 & \lambda^\alpha N_s \end{bmatrix} \begin{bmatrix} Q_s^* \\ \widetilde{Q_s}^* \end{bmatrix} \begin{bmatrix} Q_s & \widetilde{Q_s} \end{bmatrix} \begin{bmatrix} x_1(t) \\ x_2(t) \end{bmatrix} \tag{55}$$

which leads to:

$$\begin{bmatrix} Q_s & \widetilde{Q}_s \end{bmatrix} \begin{bmatrix} G_s & D_s \\ 0 & N_s \end{bmatrix} \begin{bmatrix} D^\alpha x_1(t) \\ D^\alpha x_2(t) \end{bmatrix} = \begin{bmatrix} Q_s & \widetilde{Q}_s \end{bmatrix} \begin{bmatrix} x_1(t) \\ x_2(t) \end{bmatrix} + \begin{bmatrix} Q_s & \widetilde{Q}_s \end{bmatrix} \begin{bmatrix} \lambda^\alpha G_s & \lambda^\alpha D_s \\ 0 & \lambda^\alpha N_s \end{bmatrix} \begin{bmatrix} x_1(t) \\ x_2(t) \end{bmatrix}. \tag{56}$$

Now, since $Q = \begin{bmatrix} Q_s & \widetilde{Q}_s \end{bmatrix}$ is invertible, we obtain:

$$\begin{bmatrix} G_s & D_s \\ 0 & N_s \end{bmatrix} \begin{bmatrix} D^\alpha x_1(t) \\ D^\alpha x_2(t) \end{bmatrix} = \left(I + \begin{bmatrix} \lambda^\alpha G_s & \lambda^\alpha D_s \\ 0 & \lambda^\alpha N_s \end{bmatrix} \right) \begin{bmatrix} x_1(t) \\ x_2(t) \end{bmatrix} \tag{57}$$

Thus, (57) yields the following coupled equations:

$$G_s D^\alpha x_1(t) = (1 + \lambda^\alpha G_s) x_1(t) + \lambda^\alpha D_s x_2(t), \tag{58}$$

$$N_s D^\alpha x_2(t) = (1 + \lambda^\alpha N_s) x_2(t). \tag{59}$$

Since G_s is invertible, (58) can be rewritten as:

$$D^\alpha x_1(t) = \left(G_s^{-1} + \lambda^\alpha I \right) x_1(t) + \lambda^\alpha G_s^{-1} D_s x_2(t). \tag{60}$$

Moreover, since N_s is a nilpotent matrix with $N_s^{d} = 0$, pre-multiplying (59) by N_s^{d-1} implies:

$$0 = N_s^d D^\alpha x_2(t) = \left(N_s^{d-1} + \lambda^\alpha N_s^d \right) x_2(t) = N_s^{d-1} x_2(t). \tag{61}$$

This implies that $N_s^{d-1} D^\alpha x_2(t) = 0$. Again, we have:

$$0 = N_s^{d-1} D^\alpha x_2(t) = \left(N_s^{d-2} + \lambda^\alpha N_s^{d-1} \right) x_2(t) = N_s^{d-2} x_2(t). \tag{62}$$

This also implies that $N_s^{d-2} D^\alpha x_2(t) = 0$. Repeating this process with decreasing powers of N_s eventually leads to $x_2(t) = 0$ for all t. Therefore, the subsystems (59) and (60), respectively, become:

$$D^\alpha x_1(t) = \left(G_s^{-1} + \lambda^\alpha I \right) x_1(t) \tag{63}$$

and

$$x_2(t) = 0. \tag{64}$$

Obviously, according to the Schur basis of (53) and from (19–21), the recursive solution of (63) is given by:

$$x(t) = Q_s E_{\alpha,1}\left(\left(G_s^{-1} + \lambda^\alpha I \right) t^\alpha \right) Q_s^* \, v = Q_s \left[\sum_{k=0}^{\infty} \frac{\left((G_s^{-1} + \lambda^\alpha I) t^\alpha \right)^k}{\Gamma(k\alpha + 1)} \right] Q_s^* v \tag{65}$$

and

$$y(t) = Cx(t). \tag{66}$$

Hence, the general solution (65) of the singular FoLTI systems with regular pencils is completely identified. $\square$

6. Illustrative Examples

To illustrate the main ideas of this work, this section includes two numerical examples to highlight the main ideas of the proposed approach of solving singular FoLTI continuous systems.

Example 1. *Consider the following singular FoLTI system, where R and L are constants, while u(t) is an input signal:*

$$LD^\alpha x_3(t) = u(t),\tag{67a}$$

$$Rx_2(t) = u(t),\tag{67b}$$

$$x_1(t) - x_2(t) - x_3(t) = 0.\tag{67c}$$

One may rewrite system (67) in the following form:

$$\begin{bmatrix} 0 & 0 & L \\ 0 & 0 & 0 \\ 0 & 0 & 0 \end{bmatrix} D^\alpha \begin{bmatrix} x_1(t) \\ x_2(t) \\ x_3(t) \end{bmatrix} = \begin{bmatrix} 0 & 0 & 0 \\ 1 & -1 & -1 \\ 0 & -R & 0 \end{bmatrix} \begin{bmatrix} x_1(t) \\ x_2(t) \\ x_3(t) \end{bmatrix} + \begin{bmatrix} 1 \\ 0 \\ 1 \end{bmatrix} u(t),\tag{68}$$

where:

$$E = \begin{bmatrix} 0 & 0 & L \\ 0 & 0 & 0 \\ 0 & 0 & 0 \end{bmatrix}, A = \begin{bmatrix} 0 & 0 & 0 \\ 1 & -1 & -1 \\ 0 & -R & 0 \end{bmatrix}, \text{and } B = \begin{bmatrix} 1 \\ 0 \\ 1 \end{bmatrix}.\tag{69}$$

Obviously, E is singular since $\det E = 0$. However, the pencil (E, A) is regular, because:

$$\det(E\lambda^\alpha - A) = \begin{vmatrix} 0 & 0 & L\lambda^\alpha \\ -1 & 1 & 1 \\ 0 & R & 0 \end{vmatrix} = -RL\lambda^\alpha \neq 0.\tag{70}$$

From (46), the solution of system (68) is obtained as follows:

$$\text{Let } P = \begin{bmatrix} \frac{1}{L} & 0 & 0 \\ 0 & 0 & \frac{-1}{R} \\ 0 & 1 & \frac{-1}{R} \end{bmatrix}, \text{ and } Q = \begin{bmatrix} 1 & 0 & 1 \\ 0 & 1 & 0 \\ 1 & 0 & 0 \end{bmatrix}; Q^{-1} = \begin{bmatrix} 1 & 0 & 1 \\ 0 & 1 & 0 \\ 1 & 0 & -1 \end{bmatrix}.\tag{71}$$

It follows that:

$$PEQ = \begin{bmatrix} 1 & 0 & 0 \\ 0 & 0 & 0 \\ 0 & 0 & 0 \end{bmatrix}, PAQ = \begin{bmatrix} 0 & 0 & 0 \\ 0 & 1 & 0 \\ 0 & 0 & 1 \end{bmatrix}, PB = \begin{bmatrix} \frac{1}{L} \\ \frac{-1}{R} \\ \frac{-1}{R} \end{bmatrix}.\tag{72}$$

From (28), system (69) can be transformed to the following form:

$$\begin{bmatrix} 1 & 0 & 0 \\ 0 & 0 & 0 \\ 0 & 0 & 0 \end{bmatrix}\begin{bmatrix} 1 & 0 & 1 \\ 0 & 1 & 0 \\ 1 & 0 & -1 \end{bmatrix} D^\alpha \begin{bmatrix} x_1(t) \\ x_2(t) \\ x_3(t) \end{bmatrix} = \begin{bmatrix} 0 & 0 & 0 \\ 0 & 1 & 0 \\ 0 & 0 & 1 \end{bmatrix}\begin{bmatrix} 1 & 0 & 1 \\ 0 & 1 & 0 \\ 1 & 0 & -1 \end{bmatrix} \begin{bmatrix} x_1(t) \\ x_2(t) \\ x_3(t) \end{bmatrix} + \begin{bmatrix} \frac{1}{L} \\ \frac{-1}{R} \\ \frac{-1}{R} \end{bmatrix} u(t).\tag{73}$$

In light of (34) and (35), further transformation of (73), respectively, yields:

$$\begin{bmatrix} 1 & 0 & 0 \\ 0 & 0 & 0 \\ 0 & 0 & 0 \end{bmatrix} D^\alpha \begin{bmatrix} w_1(t) \\ w_2(t) \end{bmatrix} = \begin{bmatrix} 0 & 0 & 0 \\ 0 & 1 & 0 \\ 0 & 0 & 1 \end{bmatrix} \begin{bmatrix} w_1(t) \\ w_2(t) \end{bmatrix} + \begin{bmatrix} H \\ K \end{bmatrix} u(t)\tag{74}$$

with

$$w(t) = \begin{bmatrix} 1 & 0 & 1 \\ 0 & 1 & 0 \\ 1 & 0 & -1 \end{bmatrix} \begin{bmatrix} x_1(t) \\ x_2(t) \\ x_3(t) \end{bmatrix}, \begin{bmatrix} H \\ K \end{bmatrix} = \begin{bmatrix} \frac{1}{L} \\ \frac{-1}{R} \\ \frac{-1}{R} \end{bmatrix}.\tag{75}$$

Thus:

$$D^\alpha w_1(t) = D^\alpha x_3(t) = \tfrac{1}{L} u(t)\ ,\tag{76a}$$

$$w_2(t) = \begin{bmatrix} x_2(t) \\ x_1(t) - x_3(t) \end{bmatrix} = - \begin{bmatrix} \frac{-1}{R} \\ \frac{-1}{R} \end{bmatrix} u(t). \tag{76b}$$

Using (40), the general solution of (67) is given by:

$$\begin{bmatrix} x_1(t) \\ x_2(t) \\ x_3(t) \end{bmatrix} = \begin{bmatrix} \frac{1}{R}u(t) + x_3(0) + \frac{1}{L\Gamma(\alpha)} \int_0^t (t-\tau)^{\alpha-1} u(\tau) d\tau \\ \frac{1}{R}u(t) \\ x_3(0) + \frac{1}{L\Gamma(\alpha)} \int_0^t (t-\tau)^{\alpha-1} u(\tau) d \end{bmatrix} \tag{77}$$

Example 2. *Consider system (47) for* $0 < \alpha \leq 1$ *where:*

$$E = \begin{bmatrix} 1 & 0 & 0 & 0 \\ 0 & 0 & 1 & 0 \\ 0 & 0 & 0 & 0 \\ 0 & 0 & 0 & 0 \end{bmatrix}, A = \begin{bmatrix} 0 & 0 & 0 & 0 \\ 1 & 0 & 0 & 0 \\ 0 & 1 & 0 & 0 \\ 0 & 0 & 0 & 1 \end{bmatrix}, C = [1 \ 0 \ 1 \ 0], \tag{78}$$

subject to the initial condition

$$x(0) = v = \begin{bmatrix} 1 \\ 1 \\ 0 \\ 1 \end{bmatrix}. \tag{79}$$

Notice that E is a singular matrix since $\det E = 0$, and the pencil (E, A) is regular, because:

$$\det(A - \lambda^\alpha E) = \begin{vmatrix} -\lambda^\alpha & 0 & 0 & 0 \\ 1 & 0 & -\lambda^\alpha & 0 \\ 0 & 1 & 0 & 0 \\ 0 & 0 & 0 & 1 \end{vmatrix} = -\lambda^{2\alpha} \neq 0. \tag{80}$$

Observe that (78) represents the parameters of a singular FoLTI regular system. If follows from (52) that:

$$E_s = (A - \lambda^\alpha E)^{-1} E = \begin{bmatrix} -\frac{1}{\lambda^\alpha} & 0 & 0 & 0 \\ 0 & 0 & 1 & 0 \\ -\frac{1}{\lambda^{2\alpha}} & 0 & -\frac{1}{\lambda^\alpha} & 0 \\ 0 & 0 & 0 & 0 \end{bmatrix}. \tag{81}$$

Thus, the Schur factorization for E_s is:

$$Q = \begin{bmatrix} 0 & 1 & 0 & 0 \\ 0 & 0 & 1 & 0 \\ 1 & 0 & 0 & 0 \\ 0 & 0 & 0 & 1 \end{bmatrix}, T = \begin{bmatrix} \frac{-1}{\lambda^\alpha} & \frac{-1}{\lambda^{2\alpha}} & 0 & 0 \\ 0 & \frac{-1}{\lambda^\alpha} & 0 & 0 \\ 0 & 0 & 0 & 0 \\ 0 & 0 & 0 & 0 \end{bmatrix}, Q^* = \begin{bmatrix} 0 & 0 & 1 & 0 \\ 1 & 0 & 0 & 0 \\ 0 & 1 & 0 & 0 \\ 0 & 0 & 0 & 1 \end{bmatrix}. \tag{82}$$

That is:

$$Q_s = \begin{bmatrix} 0 & 1 \\ 0 & 0 \\ 1 & 0 \\ 0 & 0 \end{bmatrix}, \widetilde{Q}_s = \begin{bmatrix} 0 & 0 \\ 1 & 0 \\ 0 & 0 \\ 0 & 1 \end{bmatrix}, G_s = \begin{bmatrix} \frac{-1}{\lambda^\alpha} & \frac{-1}{\lambda^{2\alpha}} \\ 0 & \frac{-1}{\lambda^\alpha} \end{bmatrix}, D_s = N_s = \begin{bmatrix} 0 & 0 \\ 0 & 0 \end{bmatrix}. \tag{83}$$

Moreover:

$$\left(G_s^{-1} + \lambda^\alpha I\right) t^\alpha = \begin{bmatrix} 0 & t^\alpha \\ 0 & 0 \end{bmatrix}. \tag{84}$$

Using (65), the general solution is given by:

$$x(t) = Q_s E_{\alpha,1}\left(\left(G_s^{-1} + \lambda^\alpha I\right)t^\alpha\right)Q_s^* v = \begin{bmatrix} 0 & 1 \\ 0 & 0 \\ 1 & 0 \\ 0 & 0 \end{bmatrix}\left[\sum_{k=0}^{\infty}\frac{\begin{bmatrix} 0 & t^\alpha \\ 0 & 0 \end{bmatrix}^k}{\Gamma(k\alpha+1)}\right]\begin{bmatrix} 0 & 0 & 1 & 0 \\ 1 & 0 & 0 & 0 \end{bmatrix}\begin{bmatrix} 1 \\ 1 \\ 0 \\ 1 \end{bmatrix}. \tag{85}$$

Now, since:

$$\begin{bmatrix} 0 & t^\alpha \\ 0 & 0 \end{bmatrix}^k = 0, k = 2,\ 3,\ 4,\ \dots \tag{86}$$

Thus, (85) reduces to:

$$x(t) = \begin{bmatrix} 0 & 0 & \frac{t^\alpha}{\Gamma(\alpha+1)} & 0 \end{bmatrix}^{\mathrm{T}} \tag{87}$$

Finally, from (77) and (87), $y(t) = Cx(t) = \frac{t^\alpha}{\Gamma(\alpha+1)}$.

7. Conclusions

The general solution of FoLTI continuous systems is introduced in the sense of the Caputo definition of fractional order derivative using the Adomian Decomposition Method (ADM). The same approach is extended to obtain the general solution of singular FoLTI continuous systems with regular pencils. This approach benefits from the structure of the canonical form of the system state matrices. Using the Schur decomposition, the system matrices were transformed to separate the static variables from the dynamic variables. Hence, a recursive technique is implemented to uniquely define the general solutions of both the dynamic and the static parts of the system. The case of singular FoLTI systems with singular pencils is left for further development.

Author Contributions: Authors had equal contributions in the paper. All authors have read and approved the final manuscript.

Acknowledgments: The authors thank the referees for their careful reading and valuable comments.

Conflicts of Interest: The authors declare no conflicts of interest.

References

1. Das, S. *Functional Fractional Calculus for System Identification and Controls*; Springer: Berlin, Germany, 2008.
2. Debnath, L. Recent applications of fractional calculus to science and engineering. *Int. J. Math. Math. Sci.* **2003**, *54*, 3413–3442. [CrossRef]
3. Oustaloup, A.; le Lay, L.; Mathieu, B. Identification of non-integer order system in the time domain. In Proceedings of the IEEE CESA'96, SMC IMACS Multiconference, Computational Engineering in Systems Application, Symposium on Control, Optimization and Supervision, Lille, France, 9–12 July 1996.
4. El-Khazali, R. Discretization of Fractional-Order Differentiators and Integrators. *IFAC Proc. Vol.* **2014**, *47*, 2016–2021. [CrossRef]
5. El-Khazali, R. On the biquadratic approximation of fractional-order Laplacian operators. *Analog Integr. Circuits Signal Process.* **2015**, *82*, 503–517. [CrossRef]
6. Ahmad, W.M.; El-Khazali, R.; Al-Assaf, Y. Stabilization of generalized fractional order chaotic systems using state feedback control. *Chaos Solitons Fractals* **2004**, *22*, 141–150. [CrossRef]
7. EL-Khazali, R. On the state space modeling of fractional systems. *IFAC Proc. Vol.* **2006**, *39*, 499–504. [CrossRef]
8. Trigeassou, J.-C.; Maamri, N. State space modeling of fractional differential equations and the initial condition problem. In Proceedings of the 6th International Multi-Conference on Systems, Signals and Devices, Djerba, Tunisia, 23–26 March 2009.
9. Podlubny, I. *Fractional Differential Equations*; Academic Press: San Diego, CA, USA, 1999.
10. Oldham, K.B.; Spanier, J. *The Fractional Calculus*; Academic Press: New York, NY, USA, 1974.

11. Ahmad, W.M.; EL-Khazali, R. Fractional-order dynamical models of love. *Chaos Solitons Fractals* **2007**, *33*, 1367–1375. [CrossRef]
12. Alsedá, L.L.; Cushing, J.M.; Elaydi, S.; Pinto, A.A. Difference Equations, Discrete Dynamical Systems and Applications. In *Proceedings in Mathematics and Statistics*; Springer: Berlin, Germany, 2016.
13. Gandomi, A.H.; Yun, G.J.; Yang, X.S.; Talatahari, S. Chaos-enhanced accelerated particle swarm optimization. *Commun. Nonlinear Sci. Numer. Simul.* **2013**, *18*, 327–340. [CrossRef]
14. Cánovas, J.S.; Muñoz-Guillermo, M. On the complexity of economic dynamics: An approach through topological entropy. *Chaos Soliton Fractals* **2017**, *103*, 163–176. [CrossRef]
15. Torres, H.R.; Queirós, S.; Morais, P.; Oliveira, B.; Fonseca, J.C.; Vilaca, J.L. Kidney segmentation in ultrasound magnetic resonance and computed tomography images: A. systematic review. *Comput. Methods Programs Biomed.* **2018**, *157*, 49–67. [CrossRef] [PubMed]
16. Jalab, H.A.; Ibrahim, R.W.; Ahmed, A. Image denoising algorithm based on the convolution of fractional tsallis entropy with the riesz fractional derivative. *Neural Comput. Appl.* **2017**, *28*, 217–223. [CrossRef]
17. Kilbas, A.A.; Srivastava, H.M.; Trujillo, J.J. *Theory and Applications of Fractional Differential Equations*; Elsevier: Amsterdam, The Netherlands, 2006.
18. Al-Zhour, Z. Efficient solutions of coupled matrix and matrix differential equations. *Intell. Control Autom.* **2012**, *3*, 176–187. [CrossRef]
19. Campbell, S. *Singular Systems of Differential Equations II*; Pitman: London, UK, 1982.
20. Kablar, N.; Lj, D. Debeljkovic, Finite-time stability of time varying linear singular systems. In Proceedings of the 37th IEEE Conference on Decision and Control, Tampa, FL, USA, 18 December 1998; pp. 3831–3836.
21. Takaba, T.; Morihira, N.; Katayama, K. A generalized Lyapunov theorem for descriptor system. *Syst. Control Lett.* **1995**, *24*, 49–51. [CrossRef]
22. Shahzad, A.; Jones, B.L.; Kerrigan, E.C.; Constantinides, G.A. An efficient algorithm for the solution of a coupled Sylvester equation appearing in descriptor systems. *Automatica* **2011**, *47*, 244–248. [CrossRef]
23. Junsheng, D.; Jianye, A.; Mingyu, X. Solution of system of fractional differential equations by Adomian decomposition method. *Appl. Math. J. Chin. Univ. Ser. B* **2007**, *22*, 7–12.
24. Gaxiola, G.; Bernal-Jaquez, R. Applying Adomian Decomposition Method to Solve Burgess Equation with a Non-linear Source. *Int. J. Appl. Comput. Math* **2017**, *3*, 213–224. [CrossRef]
25. Dassios, I. Optimal Solutions for non-consistent Singular Linear Systems of Fractional Nabla Difference Equations. *Circuits Syst. Signal Process.* **2015**, *34*, 1769–1797. [CrossRef]
26. Dassios, I.; Baleanu, D.I. Duality of singular linear systems of fractional nabla difference equations. *Appl. Math. Model.* **2015**, *39*, 4180–4195. [CrossRef]
27. Dassios, I.; Baleanu, D.; Kalogeropoulos, G. On non-homogeneous singular systems of fractional nabla difference equations. *Appl. Math. Comput.* **2014**, *227*, 112–131. [CrossRef]
28. Ibnbadis, A. *Contribution to Analysis and Control of Linear Singular Fractional-Order Systems*; Ministry of Higher Education and Scientific Research: Algiers, Algeria, 2017.
29. Dai, L. *Lecture Notes in Control and Information Sciences, The Theory of Matrices*; Springer: Berlin, Germany; New York, NY, USA, 1989; Volume 2.
30. Kailath, T. *Linear Systems, Information and Systems Sciences Series*; Prentice Hall: Englewood Cliffs, NJ, USA, 1980.
31. Kagstrom, B. A perturbation analysis of the generalized Sylvester equation. *SIAM J. Matrix Anal. Appl.* **1994**, *15*, 1045–1060. [CrossRef]
32. Caputo, M. Linear models of dissipation whose Q is almost frequency independent: Part II. *Geophys. J. Int.* **1967**, *13*, 529–539. [CrossRef]
33. Kaczorek, T.; Rogowski, K. *Fractional Linear Systems and Electrical Circuits*; Springer International Publishing: Cham, Switzerland, 2015.
34. Al-Zhour, Z. The general solutions of singular and non-singular matrix fractional time-varying descriptor systems with constant coefficient matrices in Caputo sense. *Alex. Eng. J.* **2016**, *55*, 1675–1681. [CrossRef]
35. Kaczorek, T. Singular fractional continuous-time and discrete-time linear systems. *Acta Mech. Automat.* **2013**, *7*, 26–33. [CrossRef]
36. Jones, B.L.; Kerrigan, E.C.; Morrison, J.F. A modeling and filtering framework for the semi-discretized Navier-Stokes equations. In Proceedings of the European Control Conference, Budapest, Hungary, 23–26 August 2009; pp. 138–143.

37. Gerdin, M. *Computation of a Canonical form for Linear Differential-Algebraic Equations*; Automatic Control Communication Systems; Linkopings Universitet: Linköping, Sweden, 2004.
38. Guglielmi, N.; Overton, M.L.; Stewart, G.W. An efficient algorithm for computing the generalized null space decomposition. *SIAM J. Matrix Anal. Appl.* **2015**, *36*, 38–54. [CrossRef]

Article

Quantifying Chaos by Various Computational Methods. Part 1: Simple Systems

Jan Awrejcewicz [1,*], Anton V. Krysko [2,3], Nikolay P. Erofeev [4], Vitalyj Dobriyan [4], Marina A. Barulina [5] and Vadim A. Krysko [4]

1   Department of Automation, Biomechanics and Mechatronics, Lodz University of Technology, 1/15 Stefanowski St., 90-924 Lodz, Poland

2   Cybernetic Institute, National Research Tomsk Polytechnic University, 30 Lenin Avenue, 634050 Tomsk, Russia; anton.krysko@gmail.com

3   Department of Applied Mathematics and Systems Analysis, Saratov State Technical University, 77 Politechnicheskaya, 410054 Saratov, Russia

4   Department of Mathematics and Modeling, Saratov State Technical University, 77 Politechnicheskaya, 410054 Saratov, Russia; erofeevnp@mail.ru (N.P.E.); dobriy88@yandex.ru (V.D.); tak@san.ru (V.A.K.)

5   Precision Mechanics and Control Institute, Russian Academy of Science, 24 Rabochaya Str., 410028 Saratov, Russia; marina@barulina.ru

*   Correspondence: jan.awrejcewicz@p.lodz.pl; Tel.: +48-42-6312-225

Received: 17 January 2018; Accepted: 1 March 2018; Published: 6 March 2018

Abstract: The aim of the paper was to analyze the given nonlinear problem by different methods of computation of the Lyapunov exponents (Wolf method, Rosenstein method, Kantz method, the method based on the modification of a neural network, and the synchronization method) for the classical problems governed by difference and differential equations (Hénon map, hyperchaotic Hénon map, logistic map, Rössler attractor, Lorenz attractor) and with the use of both Fourier spectra and Gauss wavelets. It has been shown that a modification of the neural network method makes it possible to compute a spectrum of Lyapunov exponents, and then to detect a transition of the system regular dynamics into chaos, hyperchaos, and others. The aim of the comparison was to evaluate the considered algorithms, study their convergence, and also identify the most suitable algorithms for specific system types and objectives. Moreover, an algorithm of calculation of the spectrum of Lyapunov exponents based on a trained neural network has been proposed. It has been proven that the developed method yields good results for different types of systems and does not require a priori knowledge of the system equations.

Keywords: Lyapunov exponents; Wolf method; Rosenstein method; Kantz method; neural network method; method of synchronization; Benettin method; Fourier spectrum; Gauss wavelets

---

1. Introduction

The first part of the present work is focused on the numerical investigation of classical dynamical systems to estimate velocity of divergence of the neighborhood trajectories with the help of a measure coupled with the Kolmogorov entropy [1] (or metrics). In reference [1], based on the mathematical results of Oseledec [2] and Pesin [3], it has been shown that the numerically imposed relations can be treated as exact/true values. The method proposed by Wolf [1] is most widely used to verify and study chaotic dynamics. However, also the Rosenstein [4] and Kantz [5] methods are often employed to estimate the largest Lyapunov exponents. The state-of-the-art of papers devoted to the theoretical background of the Lyapunov exponents and methods of their computations has been carried out by Awrejcewicz et al. [6]. In particular, the method of the choice of an embedding dimension has been described. The method of the correlating dimension, the false nearest neighbor method and the method of gamma-test have been presented based on the Hénon and Lorenz attractors. In particular, the occurrence of high computational difficulties has been observed in the case of the Wolf method and its marginally successful employment to small values of the studied data.

To avoid the abovementioned drawbacks, a novel neural network-based algorithm to estimate the largest Lyapunov exponents by considering only one coordinate has been proposed. In reference [6] have reported the neural network algorithm for computation of a full spectrum of Lyapunov exponents. A comparison of the results obtained by Golovko with the exact values of the Lyapunov exponents of the Lorenz and Hénon systems have exhibited small errors.

In References [7,8], the method of largest Lyapunov exponent computation using the synchronization phenomena of identical systems has been proposed. A few types of coupling have been studied, depending on the type of the considered system. It has been pointed out that large computation time is required to achieve full synchronization.

The method proposed in References [9,10] is particularly suitable to study chaotic dynamics of continuous mechanical systems. It should be emphasized that, owing to the results published by the authors of the present paper, the analysis of nonlinear dynamics based on the estimation of the Lyapunov exponents yields a conclusion that the mentioned problems have not been satisfactorily solved yet [1,4,5,9,10].

More recently, Vallejo and Sanjuan [11,12] have studied the predictability of orbits in coupled systems by means of finite-time Lyapunov exponents. This approach has allowed them to estimate how close the computed chaotic orbits are to the real/true orbits, being characterized by the systems shadowing properties.

In the present paper, classical systems (Hénon map [13], hyperchaotic Hénon map [14], logistic map [15], Rössler attractor [16], and Lorenz attractor [17]) were analyzed with Gauss wavelets [18], Fourier spectra and Poincaré maps of a chaotic attractor [19–21].

It is known that the fundamental property of chaos is the existence of strong sensitivity to a change of the initial conditions. The definition of chaos, given first by Devaney in 1989 [22], includes three fundamental parts. In addition to sensitivity to the variation of the initial conditions, a condition of mixing, known also as the transitivity condition and the regularity condition, measured by the density of the periodic points or classical notion of periodicity is also included. In 1992, Banks et al. [23] proved that the condition of sensitivity to the initial condition can be neglected, i.e., conditions of transitivity and periodicity imply the sensitivity condition.

Knudsen [24] has defined chaos as a function given on a bounded metric space which has a dense orbit and essentially depends on initial conditions.

Owing to the definition proposed by Gulick [25], chaos exists when either there is essential dependence on the initial conditions or a chaotic function has positive Lyapunov exponents in each point of the space and which does not eventually tend to a periodic orbit. This definition has been also employed in the present work.

2. Lyapunov Exponents

2.1. The Largest Lyapunov Exponent

The following dynamical system was considered:

$$\dot{x} = f(x) \tag{1}$$

where x stands for the N-dimensional state vector.

Two closed phase points x_1 and x_2 were chosen (in the phase space). They stand for the origins of the trajectories $(x_1(t)$ and $x_2(t))$. The change in the distance d between two corresponding points of these trajectories under evolution of system (1) can be monitored by:

$$d(t) = \left|\vec{\varepsilon}(t)\right| = |x_2(t) - x_1(t)| \tag{2}$$

If the dynamics of system (1) is chaotic, $d(t)$ increases exponentially in time, i.e.,:

$$d(t) \approx d(0)e^{kt} \tag{3}$$

This yields the average velocity of the exponential divergence of the trajectories:

$$k \approx \frac{\ln\left[\frac{d(t)}{d(0)}\right]}{t} \tag{4}$$

or more precisely:

$$k = \lim_{\substack{d(0) \to 0 \\ t \to \infty}} \frac{\ln[d(t)/d(0)]}{t} \tag{5}$$

The quantity h is known as the Kolmogorov-Sinai entropy (KS-entropy). Employing the KS-entropy, one can define the studied process, i.e., quantify if the process is regular or chaotic. In particular, if the system dynamics is periodic or quasi-periodic, the distance $d(t)$ is not inversed in time and the KS-entropy is equal to zero ($h = 0$). If the system dynamics tends to a stable fixed point $d(t) \to 0$, then $h < 0$. Contrarily, KS-entropy is positive ($h > 0$) if one deals with chaotic dynamics. KS-entropy is the maximum characteristic Lyapunov exponent that enables one to follow velocity of information lost with respect to the initial system state.

2.2. Results

The spectrum of Lyapunov exponents makes it possible to qualitatively quantify a local property with respect to the stability of an attractor. Consider a phase trajectory $x(t)$ of the dynamical system (1), starting from the point $x(0)$ as well as its neighborhood trajectory $x_1(t)$ as follows:

$$x_1(t) = x(t) + \vec{\varepsilon}(t) \tag{6}$$

The following function can be constructed:

$$\lambda\left[\vec{\varepsilon}(0)\right] = \lim_{t \to \infty} \frac{\ln\left[\frac{|\vec{\varepsilon}(t)|}{|\vec{\varepsilon}(0)|}\right]}{t} \tag{7}$$

which is defined on the vector of initial displacement $\vec{\varepsilon}(0)$ such that $\left|\vec{\varepsilon}(0)\right| = \varepsilon$, where $\varepsilon \to 0$.

All possible rotations of the initial displacements vector with respect to n directions of the N-dimensional phase space of the Function (7) will suffer the jump-like changes in the finite series of the values $\lambda_1, \lambda_2, \lambda_3, \ldots, \lambda_n$. These values of the function λ are called Lyapunov exponents (LEs). Positive/negative values of LEs can be viewed as a measure of the averaged exponential divergence/convergence of the neighborhood trajectories.

A sum of LEs stands for an averaged divergence of the phase trajectories flow. In the case of a dissipative system, i.e., a system possessing an attractor, this sum is always negative. As numerical case studies show, in some dissipative systems the LEs are invariant with respect to the chosen initial conditions. Hence, a spectrum of LEs can be understood as the property of an attractor.

Usually, LEs are presented in a sequence of LE values in decreasing order. For instance, symbols $(+, 0, -)$ mean that for the analyzed attractor, there is one direction in a 3D space, where exponential stretching is exhibited, the second direction indicates neutral stability, and the third one—exponential compression. It should be noted that all attractors different from stable stationary points always have at least one LE equal to zero (in average sense, all points of a trajectory are bounded by a compact manifold and they cannot exhibit divergence or converge).

In what follows, relationships between the Lyapunov exponents and the properties and types of attractors are illustrated and discussed:

(1) $n = 1$. In this case only a stable fixed point can be an attractor (node or focus). There exists one negative LE denoted by $\lambda_1 = (-)$,

(2) $n = 2$. In 2D systems, there are two types of attractors: stable fixed points and limit cycles. The corresponding LEs follow:

- $(\lambda_1, \lambda_2) = (-, -)$—stable fixed/fixed point;
- $(\lambda_1, \lambda_2) = (0, -)$—stable limit cycle (one exponent is equal to zero).

(3) $n = 3$. In 3D phase space, there exist four types of attractors: stable points, limit cycles, 2D tori and strange attractors. The following set of LEs characterizes possible dynamical situations to be met:

- $(\lambda_1, \lambda_2, \lambda_3) = (-, -, -)$—stable fixed point;
- $(\lambda_1, \lambda_2, \lambda_3) = (0, -, -)$—stable limit cycle;
- $(\lambda_1, \lambda_2, \lambda_3) = (0, 0, -)$—stable 2D tori;
- $(\lambda_1, \lambda_2, \lambda_3) = (+, 0, -)$—strange attractor.

In the majority of the studied problems, it is impossible to give an analytical definition of LEs, since the analytical solution to the governing differential equations would have to be known. However, there exist reliable algorithms to find all Lyapunov exponents numerically.

3. Methods of Analysis of Lyapunov Exponents

3.1. Benettin Method

We began with the numerical investigation of the Kolmogorov entropy of the Hénon-Heiles model. Numerical computations were carried out with accuracy up to 14 digits by means of employing the so-called method of central points. Observe that a similar method has been used in reference [26].

Based on the Lyapunov exponents, the ergodic properties of dissipative dynamical systems with a few degrees of freedom were numerically studied with the Lorenz system. The system exhibited the exponents spectrum of the (+, 0, −) type, and the exponents had the same values for the orbits beginning from an arbitrary point on the attractor. It means that the ergodic property of a general dynamical system can be quantified by a spectrum of the characteristic Lyapunov exponents. Below, a brief description of the used method is presented.

Let a point x_0 belong to the attractor A of a dynamical system. An evolution trajectory of the point x_0 is referred to as a real/true trajectory. A positive quantity ε, being significantly less than the attractor dimension, is chosen. Furthermore, an arbitrary perturbed point $\tilde{x}_0$ is chosen in a way to satisfy $\|\tilde{x}_0 - x_0\| = \varepsilon$. The evolution of points x_0 and $\tilde{x}_0$ is considered in a short time interval T, and new points are denoted by x_1 and $\tilde{x}_1$, respectively. A vector $\Delta x_1 = \tilde{x}_1 - x_1$ is called the perturbation vector. The first estimate of the exponent is found with the use of the following formula

$$\tilde{\lambda}_1 = \frac{1}{T} \ln \frac{\|\Delta x_1\|}{\varepsilon}. \tag{8}$$

The time interval T is chosen in a way to keep the amplitude of perturbation less than the linear dimensions of the phase space nonhomogenity and the attractor dimension. The normalized perturbation vector $\Delta x_1' = \varepsilon \Delta x_1 / \|\Delta x_1\|$ is taken, and a new perturbed point $\tilde{x}_1' = x_1 + \Delta x_1'$ is defined. Finally, the so far described procedure is implemented taking into account x_1 and $\tilde{x}_1$ instead of x_0 and $\tilde{x}_0$, respectively.

After repeating this procedure M times, λ is defined as an arithmetic average of the estimates $\tilde{\lambda}_l$ obtained on each computational step:

$$\lambda \cong \frac{1}{M}\sum_{i=1}^{M}\tilde{\lambda}_{l} = \frac{1}{M}\sum_{i=1}^{M}\frac{1}{T}\ln\frac{||\Delta x_{i}||}{\varepsilon} = \frac{1}{MT}\sum_{i=1}^{M}\ln\frac{||\Delta x_{i}||}{\varepsilon}. \tag{9}$$

In order to achieve a higher estimate, one can take large M and carry out computations for a different initial point x_0. This method can be used when the equations governing the system evolution are known. It should be noted, however, that these equations are usually unknown for the experimental data.

To compute the Lyapunov spectrum numerically, one can use another approach generalizing the Benettin's algorithm. In general, it is necessary to follow a few trajectories of the perturbed points instead of only one, fundamental trajectory (the number of perturbed trajectories is equal to the dimension of the phase space). For this purpose, a numerical approach based on derivation of the dynamic equations in variations can be used [27]. Since the largest LE plays a crucial role in the evolution of all perturbed trajectories, it is necessary to carry out orthogonalization of the perturbation vectors on each step of the algorithm. In what follows, a procedure of numerical estimation of the Lyapunov spectrum of a dynamical system is briefly described. To simplify, the considerations are limited to 3D systems.

Let r_0 stand for a point of the chaotic attractor and ε be a fixed positive number, small in comparison to linear dimensions of the attractor. The points x_0, y_0 and z_0 are chosen so that the perturbation vectors $\Delta x_0 = x_0 - r_0$, $\Delta y_0 = y_0 - r_0$, $\Delta z_0 = z_0 - r_0$ have the length ε and are mutually orthogonal. After a certain small time interval T, the points r_0, x_0, y_0 and z_0 are transformed into points r_1, x_1, y_1 and z_1, respectively. Then, new perturbation vectors $\Delta x_1 = x_1 - r_1$, $\Delta y_1 = y_1 - r_1$, $\Delta z_1 = z_1 - r_1$ are considered. The orthogonlization using the well-known (in linear algebra) Gramm–Schmidt method is carried out. After this step, the obtained vectors of perturbation $\Delta x_1'', \Delta y_1'', \Delta z_1''$ become orthonormalized, i.e., they are mutually orthogonal and have the unit length. Then, the renormalization of the perturbation vectors is carried out again to get lengths of the vectors in terms of the magnitude ε:

$$\Delta x_1''' = \Delta x_1'' \times \varepsilon, \quad \Delta y_1''' = \Delta y_1'' \times \varepsilon, \quad \Delta z_1''' = \Delta z_1'' \times \varepsilon \tag{10}$$

We take the following perturbed points:

$$x_1' = x_1 + \Delta x_1''', \quad y_1' = y_1 + \Delta y_1''', \quad z_1' = z_1 + \Delta z_1''' \tag{11}$$

Next, the process is repeated, i.e., instead of the points r_0, x_0, y_0 and z_0, the points r_1, x_1', y_1' and z_1' are taken into account, respectively.

Repeating the so far described procedure M times, one finds:

$$S_1 = \sum_{k=1}^{M}\ln||\Delta x_k'||, \quad S_2 = \sum_{k=1}^{M}\ln||\Delta y_k'||, \quad S_3 = \sum_{k=1}^{M}\ln||\Delta z_k'||. \tag{12}$$

Then, a spectrum $\Lambda = \{\lambda_1, \lambda_2, \lambda_3\}$ of LEs can be found by the following formulas:

$$\lambda_i = \frac{S_i}{MT}, \quad i = 1, 2, 3 \tag{13}$$

In this method, the choice of time interval T is crucial. If one takes too large time interval T, then all perturbed trajectories are inclined in the direction corresponding to the maximum LE, and hence the obtained results are not reliable.

3.2. Wolf Method

In Reference [1], a novel algorithm to find nonnegative Lyapunov exponents by using a time series has been proposed. It has been illustrated that the Lyapunov exponents are associated with either exponential divergence or convergence of the neighborhood orbits in the considered phase space. In general, the method is applicable only when analytical governing equations are known, and it is based on tracing the large time-consuming increase in the number of elements in a small volume of an attractor.

We defined a Lyapunov exponent and a spectrum of Lyapunov exponents, and then illustrated how the system dynamics depends on the number of exponents with different signs in the spectrum. Our approach included reconstruction of an attractor and investigation of orbital divergence on the possibly smallest distances using the approximate Gramm–Schmidt orthogonalization procedure in the reconstructed phase. In order to estimate the largest Lyapunov exponent, a long trace of time evolution of the chosen pair of the neighborhood orbits was carried out. Note that a particular attention should be paid, since the reconstructed attractor may contain points belonging to different attractors.

Two versions of the method are proposed. The first one includes the so-called fixed evolution time, where the time interval associated with the change of the points is fixed.

The main idea of the proposed method is that the largest Lyapunov exponent is computed based on one time series and used when the equations describing the system evolution are unknown and when it is impossible to measure all remaining phase coordinates.

Consider a time series $x(t)$, $t = 1, \ldots, N$ of one coordinate of a chaotic process measured in equal time intervals. The method of mutual information allows one to define the time delay τ, whereas the method of false neighbors yields the dimension of the embedded space m. As a result of the reconstruction, one gets a set of points of the space R^m:

$$x_i = (x(i), x(i - \tau), \ldots, x(i - (m - 1)\tau)) = (x_1(i), x_2(i), \ldots, x_m(i)), \tag{14}$$

where $i = ((m - 1)\tau + 1), \ldots, N$.

We take a point from the series (3) and denote it by x_0. In the series (3), one can find a point $\tilde{x}_0$, where the relation $||\tilde{x}_0 - x_0|| = \varepsilon_0 < \varepsilon$ holds, and where ε is a fixed quantity, essentially less than the dimension of the reconstructed attractor. It is required that the points x_0 and $\tilde{x}_0$ are separated in time. Then, time evolution of these points is observed on the reconstructed attractor until the distance between points achieves $\varepsilon_{\max}$. The new points are denoted by x_1 and $\tilde{x}_1$, the distance is ε_0', and the associate interval of time evolution is denoted by T_1.

After that, we again consider the Sequence (14) the find the point $\tilde{x}_1'$, located close to x_1, where $||\tilde{x}_1' - x_1|| = \varepsilon_1 < \varepsilon$ holds. Vectors $\tilde{x}_1 - x_1$ and $\tilde{x}_1' - x_1$ should possibly have the same direction. Then, the procedure is repeated for points x_1 and $\tilde{x}_1'$.

By repeating the above procedure M times, the largest Lyapunov exponent is estimated:

$$\lambda \cong \sum_{k=0}^{M-1} \ln(\varepsilon_k'/\varepsilon_k) / \sum_{k=1}^{M} T_k. \tag{15}$$

This method has been employed in the present research to test the accuracy of results by using the classical and known spectra of the Lyapunov exponents of the Hénon map, Rössler equations, chaos and hyperchaos exhibited by the Lorenz system, and McKay-Glass equation [28]. In addition, it has been also employed to study the Belousov–Zhabotinsky reaction [29] and the Couette-Taylor flow [30].

Wolf et al. [1] have pointed out certain restrictions on the choice of the embedding dimension and the time required for the attractor reconstruction to achieve the most accurate estimates of the Lyapunov exponents. Using the Rössler attractor [16] and the Belousov–Zhabotynskiy reaction [29], the authors have demonstrated the effects of the time change during the attractor reconstruction, the time of evolution of the system between steps of the time change, the maximum length of the replacement vector and the minimum length of the exchange vector on the values of the estimated largest Lyapunov exponent. Furthermore, it has been shown that variation (between 0.5 and 1.5) of the time of the system evolution leads to reliable estimates of the studied three chaotic attractors. Also, some data requirements that make it possible to obtain the most accurate estimate of the Lyapunov exponent, such as the use of small length scale data as well as some restrictions on the presence of noisy perturbations in the data (static and dynamic), have been discussed.

The proposed algorithms can be used to detect chaos as well as to compute its parameters also for the experimental data with a few positive exponents. Furthermore, numerical studies have presented the topological complexity of chaos (the Lorenz attractor) and have shown that the deterministic chaos can be distinguished from white noise (the Belousov–Zhabotinsky reaction).

3.3. Rosenstein Method

Despite this method is simple in realization in comparison to the previous ones and it is characterized by high computational speed, it does not directly yield λ_1, but rather the function:

$$y(i, \Delta t) = \frac{1}{\Delta t} \langle \ln d_j(i) \rangle, d_j(i) = \min_{x_j} ||x_j - x'_j||, \tag{16}$$

where x_j is a given point, and x'_j denotes its neighbor.

The algorithm is based on the relationship between d_j and the Lyapunov exponents: $d_j(i) \approx e^{\lambda_1(i\Delta t)}$. The largest Lyapunov exponent is computed by estimating the inclination of the most linear part of the function. It should be mentioned, however, that finding this linear part does not belong to easy tasks.

3.4. Kantz Method

The algorithm proposed by Kantz [5] computes the LLE by searching all neighbors in vicinity of the reference trajectory and estimates the average distance between neighbors and the reference trajectory as a function of time (or a relative time multiplied by the data sampling frequency). The algorithm is based on the following formula:

$$S(\tau) = \frac{1}{T} \sum_{t=1}^{T} \ln \left(\frac{1}{|U_t|} \sum_{i \in U_t} |x_{t+\tau} - x_{i+\tau}| \right) \tag{17}$$

where x_t stands for an arbitrary signal point; U_t is a neighborhood of x_t; x_i is a neighbor of x_t; τ—relative time multiplied by the sampling frequency; T—sample size; $S(\tau)$—stretching factor in the region of a linear growth indicating a curve whose slope is equal to LE, i.e., $e^{\lambda\tau} \propto e^{S(\tau)}$. However, the assumption of a linear growth introduces new errors. Despite the fact that the method is useful and accurate for systems with known LEs, the choice of parameters and the region where the mentioned linear growth occurs is, in practice, arbitrary.

The method yields correct results if the value of the Lyapunov exponent is known a priori, and hence the space with the tangent equal to that value can be chosen.

3.5. Computation of LLE Based on Synchronization of Nonnegative Feedback

In reference [7], the method of LLE computation based on synchronization of coupled identical systems has been proposed. The following k-dimensional discrete system:

$$y'_i = f(y_i) \tag{18}$$

has been considered, where $y \in \mathbb{R}^k$, $i \in (1, 2, \ldots, k)$. The supplemental system has been proposed in the following way:

$$\begin{aligned} x'_i &= f(y_i + \Delta y_i) \\ y'_i &= f(y_i) \\ \Delta y'_i &= [f(y_i + \Delta y_i) - f(y_i)] \exp(-p) \end{aligned} \tag{19}$$

where x, y, $\Delta y \in \mathbb{R}^k$. Evolution of k-dimensional system is governed by k of LLEs. Consequently, synchronization of the perturbed and nonperturbed systems (19) is guaranteed by the following inequality:

$$p > \lambda_{\max} \tag{20}$$

where $\lambda_{\max}$ stands for LLEs of the studied systems (18).

Figure 1 shows synchronization between perturbed (first equation of (19)) and nonperturbed (second equation of (19)) systems for alogistic map. The synchronization starts at p equal to λ, and this value represents the largest Lyapunov exponent of the system.

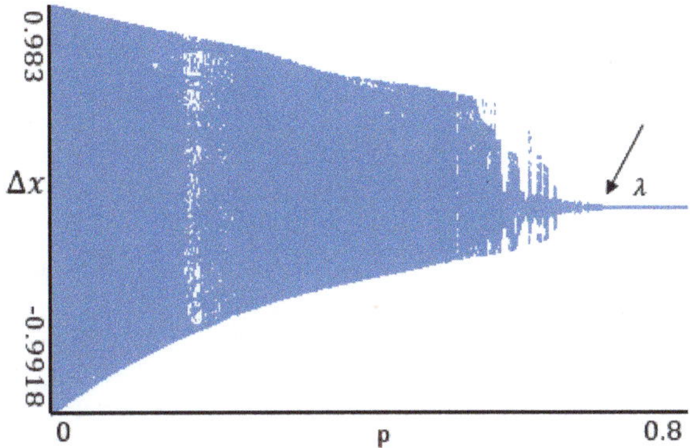

Figure 1. Synchronization of perturbed and nonperturbed systems in the case of a logistic map (λ points to the largest Lyapunov exponent value).

In reference [8], systems with excitations have been studied. The authors have proposed the following way of coupling of identical systems:

$$\begin{aligned}
\dot{x} &= f(x) \\
\dot{y} &= f(y) + d(x - y)
\end{aligned} \tag{21}$$

The application of this approach is limited to the systems with known equations of evolutions, and the way of introducing the coupling of two identical systems depends on the type of the considered system.

3.6. Jacobi Method

This method has been proposed in references [31,32]. The main idea is to use an algorithm, the scheme of which is illustrated in Figure 2. A sphere of small radius ε is taken. After a few iterations m, a certain operator T^m transforms this sphere into an ellipsoid having $a_1, \ldots, a_p$ half-axes. The sphere is stretched along the axes $a_1, \ldots, a_s > \varepsilon$, where s is the number of positive LEs. For sufficiently small ε, the operator T^m is close to the sum of the shear operator and the linear operator A. The LLEs are computed as averaged eigenvalues of the operator A on the whole attractor.

A vector ς_j is chosen, and a set $\{\varsigma_{k_i}\}$ $(i = 1, \ldots, N)$ of i-th neighborhood vectors is found. The following set of vectors $y_i \equiv \varsigma_{k_i} - \varsigma_j$, where $\|y_i\| \leq \varepsilon$, is taken. After m successive iterations, the operator T^m transforms the vector ς_j into ς_{j+m}, and the vector ς_{k_i} into $\varsigma_{k_{i+m}}$. Eventually, the vectors y_i are transformed into

$$y_{i+m} = \varsigma_{k_{i+m}} - \varsigma_{j+m}$$

Assuming that the radius ε is sufficiently small, one can introduce the operator A_j as follows

$$y_{i+m} = A_j y_i$$

The operator A_j describes the system in variations. To estimate the operator A, the least-square method can be employed:

$$\min_{A_j} S = \min_{A_j} \frac{1}{N} \sum_{i=0}^{N} \left(y_{i+m} - A_j y_i \right)^2$$

This yields the following system of equations of the dimension $n \times n$:

$$A_j V = C, \ (V)_{kl} = \frac{1}{N} \sum_{i=1}^{N} y_i^k y_i^l$$

$$(C)_{kl} = \frac{1}{N} \sum_{i=1}^{N} y_{i+m}^k y_i^l$$

where V, C are the matrices of the dimension $n \times n$, y_i^k stands for the k-th component of vector y_i, and y_{i+m}^k is the k-th component of the vector y_{i+m}. If A is a solution of the equations, then the LEs can be found in the following way

$$\lambda_i = \lim_{n \to \infty} \frac{1}{n\tau} \sum_{j=1}^{n} \ln A_j e_i^j$$

where $\{e_j\}$ is a set of basic vectors in a tangent space ς_j.

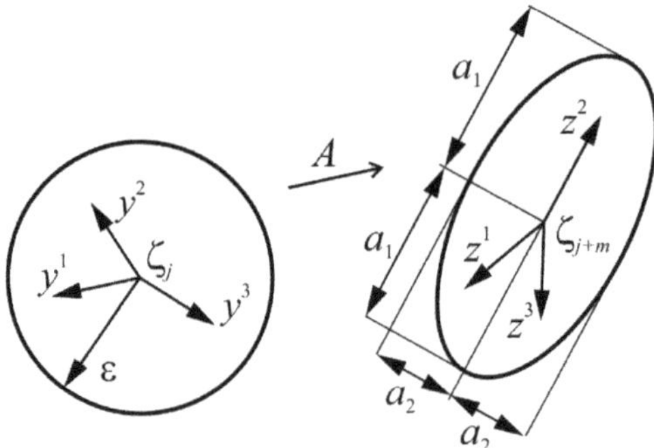

Figure 2. Transformation of a sphere of initial states into a counterpart ellipsoid during the system evolution.

The algorithm can be realized in a way similar to the computation of LEs of the ODEs given analytically.

Let us choose an arbitrary basis $\{e^s\}$ and then follow the changes in the length of the vector $A_j e^s$. As the vectors $A_j e^s$ grow and their orientations change, it is necessary to perform their orthogonalization and normalization by using, for example, the Gramm–Schmidt procedure. The procedure is then repeated for the new basis.

The mentioned method allows one to estimate a spectrum of nonnegative LEs. However, it has a serious disadvantage—it is highly sensitive to noise and errors.

3.7. Modification of the Neural Network Method

We have proposed a novel counterpart method to compute LEs based on a modification of the neural network method (see Figure 3).

A single-layer feed forward neural network presented in Figure 3 has multiple input neurons, a layer of hidden neurons and one output neuron. The following notation is employed: a_{ij}—weight of the connection between the i-th input neuron and the j-th hidden neuron; b_i—weight of connection between the i-th hidden neuron and the output neuron. To realize the neural network algorithm, the following criteria were taken into account:

(i) the network is sensitive to the input information (information is given in the form of real numbers);

(ii) the network is self-organizing, i.e., it yields the output space of solutions only based on the inputs;

(iii) the neural network is a network of straight distribution (all connections are directed from input neurons to output neurons);

(iv) owing to the synapses tuning, the network exhibits dynamic couplings (in the learning process, the tuning of the synaptic coupling takes place ($dW/dt \neq 0$), where W stands for the weighted coefficients of the network).

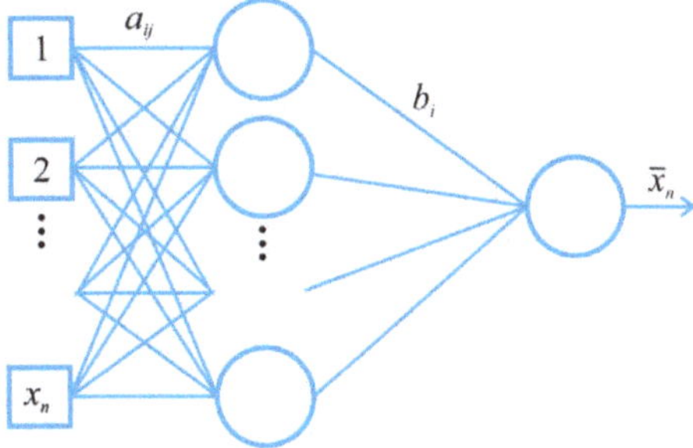

Figure 3. Single-layer feed forward neural network, which consists of input neurons, a layer of hidden neurons and one output neuron.

The hidden layer of neurons contains the hyperbolic tangent, which plays a role of an activation function (Figure 4). A derivative of the hyperbolic tangent is described by a quadratic function, as it is in the case of a logistic function. However, contrarily to the logistic function, the space of the values of the hyperbolic tangent falls within the interval $(-1; 1)$. This results in higher convergence in comparison to the standard logistic function.

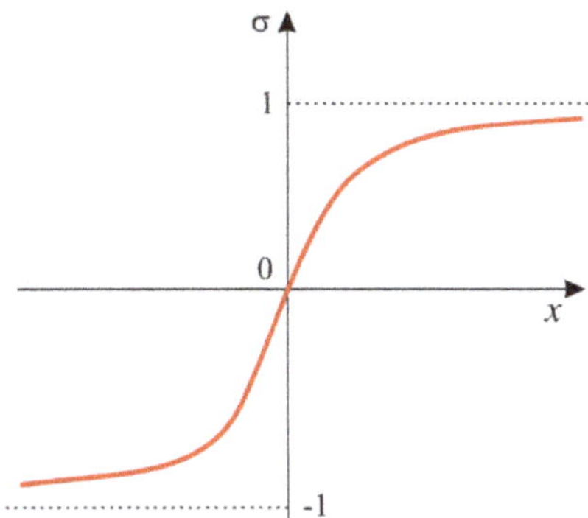

Figure 4. Transition function.

Prognosis of $\hat{x}_k$ of a scalar time series x_k is made by employing the following formula

$$\hat{x}_k = \sum_{i=1}^{n} b_i \tanh\left(a_{i0} + \sum_{j=1}^{d} a_{ij} x_{k-j} \right) \tag{22}$$

where n stands for the number of neurons, d is the number of the searched LE, a_{ij} stands for the $n \times (d+1)$ matrix of coefficients, and b_i is the vector of the length n. The matrix a_{ij} contains the coupling forces with respect to the network input, the vector b_i is used to control the input of each neuron to the network output, whereas the vector a_{i0} is used for relatively simple learning based on data with nonzero averaged value.

Weights a and b are chosen in a probabilistic way, and the dimension of the searched solution is decreased in the process of learning. The associated Gaussian is chosen in a way to have initial standard distribution 2^{-j}, centered with respect to zero in order to promote the most recent time delays (small values of j) in the phase space. The coupling forces are chosen in a way to minimize the averaged one step mean square error of a forecast:

$$e = \frac{\sum_{k=d+1}^{c} (\hat{x}_k - x_k)^2}{c - d} \tag{23}$$

During the training of the network, sensitivity of the output is defined by computing partial derivatives of all averaged points of the time series in each time step x_{k-j}:

$$\hat{S}(j) = \frac{1}{c-j} \sum_{k=j+1}^{c} \left| \frac{\partial \hat{x}_k}{\partial x_{k-j}} \right| \tag{24}$$

In the case of the network given by (22), the partial derivatives have the following form:

$$\frac{\partial \hat{x}_k}{\partial x_{k-j}} = \sum_{i=1}^{n} a_{ij} b_i \sec \mathrm{h}^2 \left(a_{i0} + \sum_{m=1}^{d} a_{im} x_{k-m} \right) \tag{25}$$

The largest value j is the optimal embedding dimension, and the key role is played by $\hat{S}(j)$ as in the false nearest neighbors method. The individual values of $\hat{S}(j)$ yield a quantitative estimate of the importance of each time step using the associated terms of the autocorrelation function or coefficients of the associated linear model.

The weights of the trained neural network are substituted to the matrix of solutions, and the input data are used to define the initial state. The computation of the spectrum is realized by employment of the generalized Benettin's algorithm based on the obtained system of equations.

4. Wavelet Methods

Gauss Wavelets

In the majority of engineering problems, the Fourier analysis is insufficient, since it deals with the averaged spectrum of the whole studied vibration signal and presents only a general picture of the signal. On the contrary, wavelets play a role of a "microscope" which allows one to observe the spectrum at each time instant, and detect births/deaths of the frequencies in time.

A wavelet transform of a 1D signal is realized with respect to a basis being usually a soliton-like function with given properties. The basis is obtained by displacement and tension/compression of a function called a wavelet.

In the present work, the Gauss wavelets, defined as derivatives of the Gauss function, were used. Higher-order derivatives have many zero moments, and hence they allow one to obtain information about higher-order features hidden in the investigated signal.

The 8th order Gauss wavelets of the of the following form were employed:

$$g_8(x) = -\left(105 - 420x^2 + 210x^4 - 28x^6 + x^8\right)\exp^{\frac{-x^2}{2}} \tag{26}$$

5. Analysis of Classical Dynamical Systems by LEs and Gauss Wavelets

In this section, simple classical systems (Figures 5–9) have been studied with emphasis put on a comparison of the LEs (Tables 1–5) obtained using the Wolf, Rosenstein, Kantz and neural network methods. The convergence of the mentioned methods, depending on the number of iteration steps, has been illustrated and discussed (Tables 6–10). The Benettin method has been used as a reference because for most systems, there are no analytically calculated spectra of Lyapunov exponents. Moreover, the Benettin method calculates Lyapunov exponents based on the system equations.

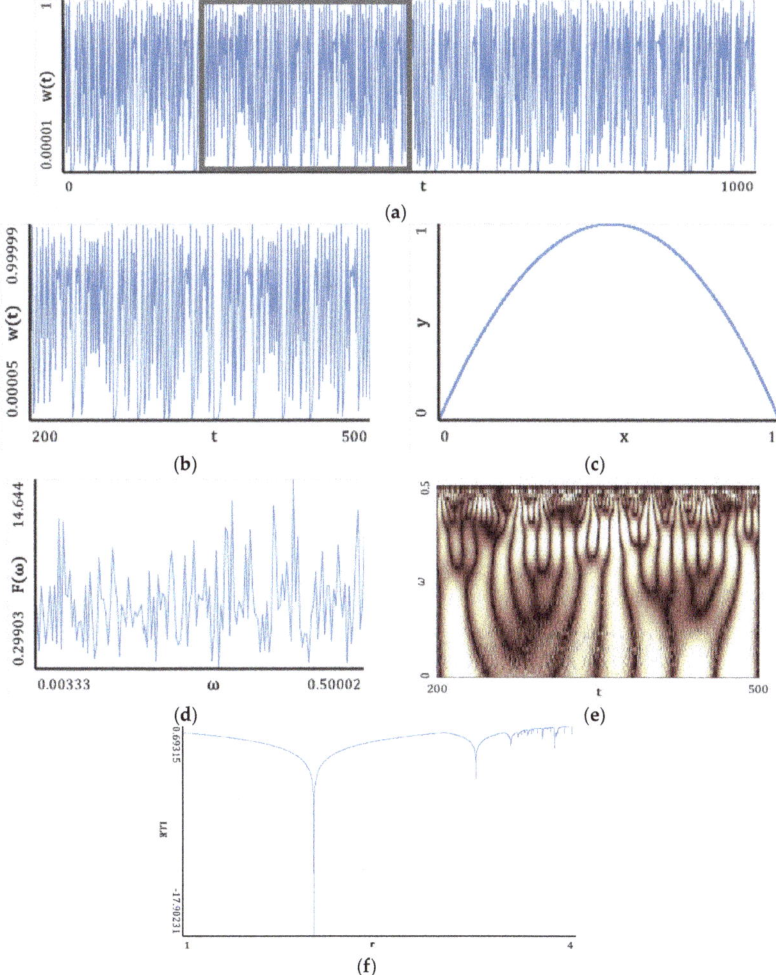

Figure 5. Nonlinear characteristics of the oscillation signal: (**a**) Time histories; (**b**) Time window; (**c**) Chaotic attractor; (**d**) Fourier frequency spectrum; (**e**) Wavelet spectrum; (**f**) Dependence of LLE on the control parameter.

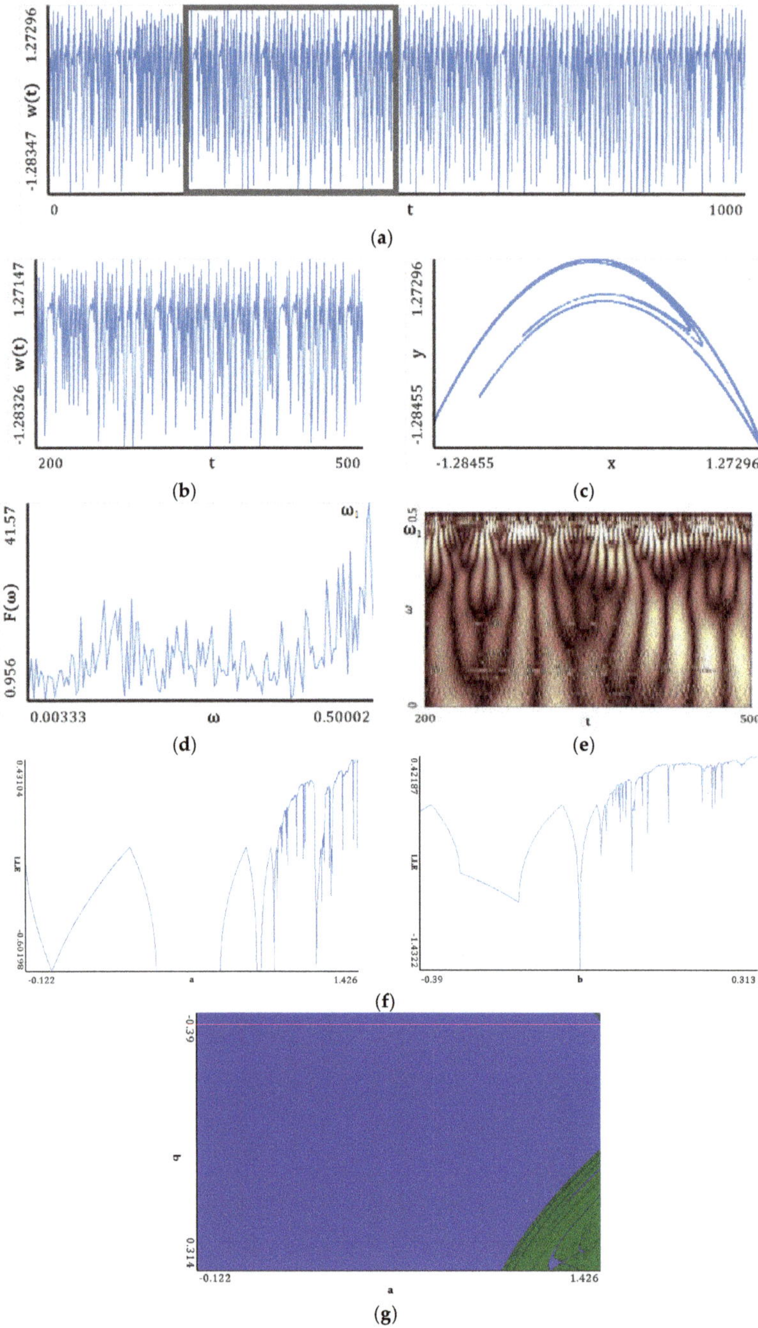

Figure 6. Characteristics of the Hénon map: (**a**) Time history; (**b**) Time window; (**c**) Chaotic attractor; (**d**) Fourier frequency spectrum; (**e**) Wavelet spectrum; (**f**) Dependence of LLE on the control parameter; (**g**) Lyapunov exponents plane (Hénon map).

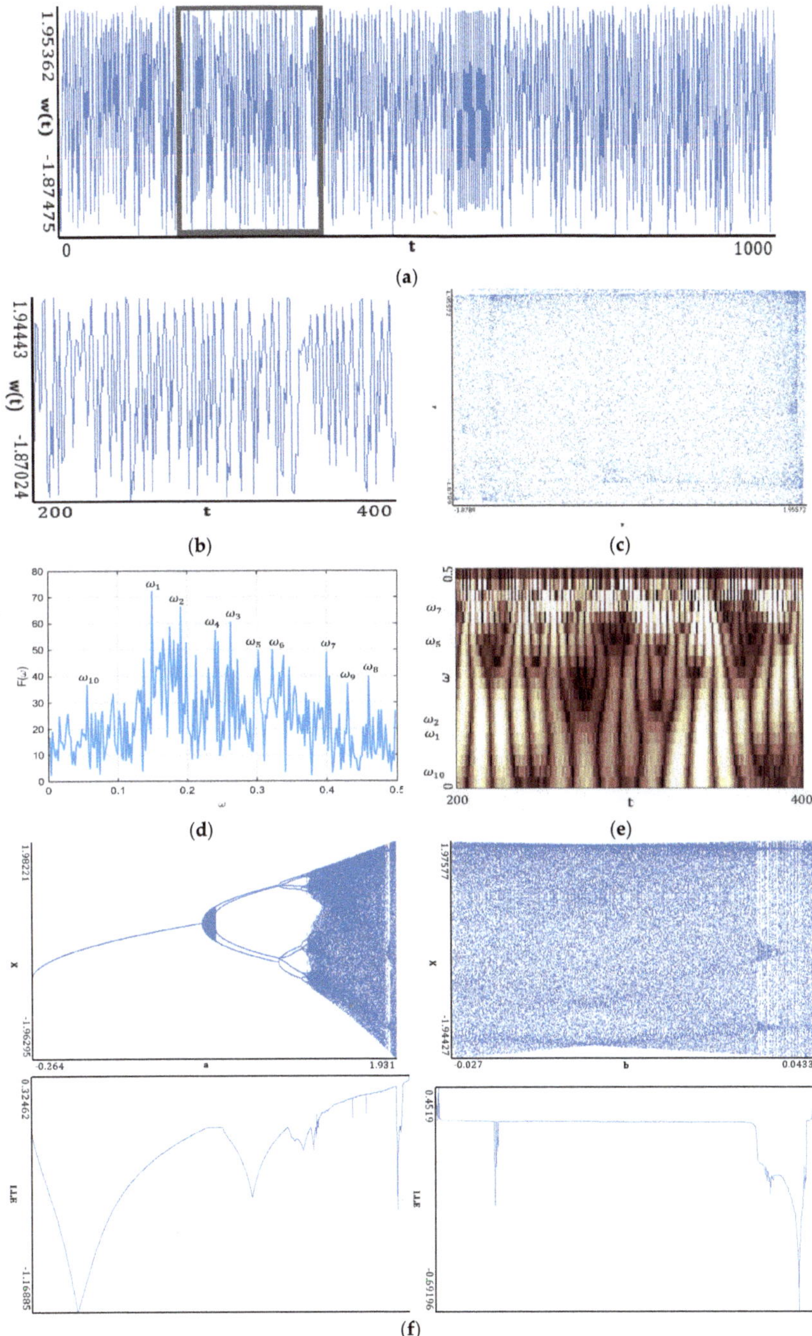

Figure 7. *Cont.*

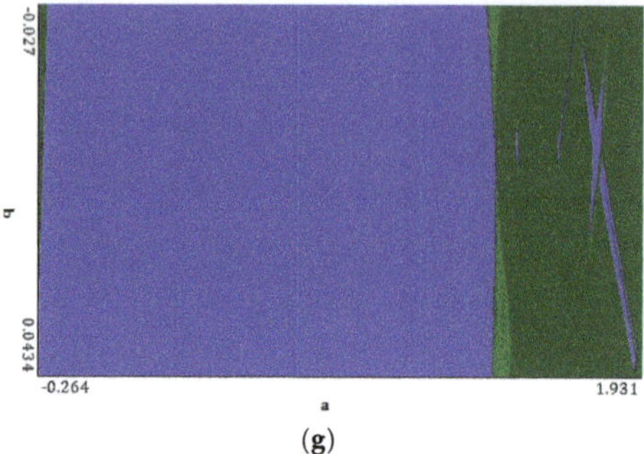

(**g**)

Figure 7. Signal characteristics: (**a**) Time history; (**b**) Time window; (**c**) Chaotic attractor; (**d**) Fourier frequency spectrum; (**e**) Wavelet spectrum; (**f**) Dependence of LLE on the control parameter; (**g**) Lyapunov exponents plane (generalized Hénon map).

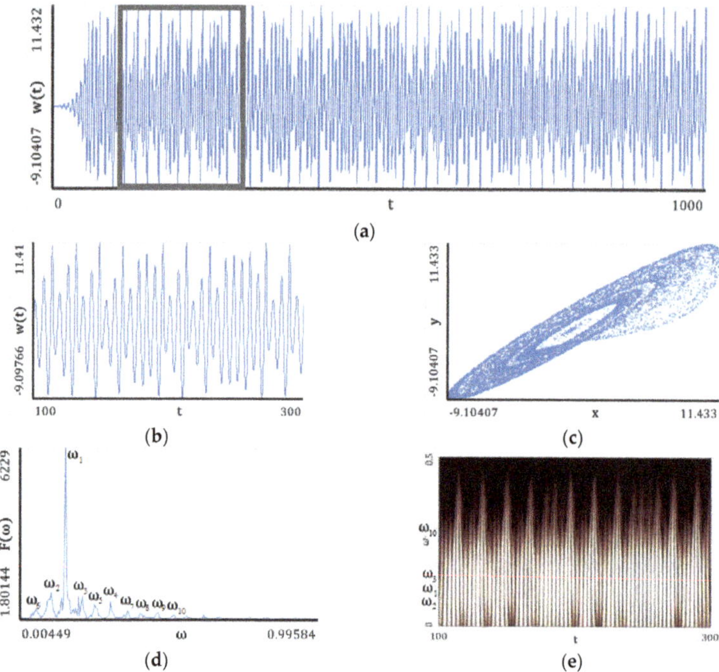

Figure 8. *Cont.*

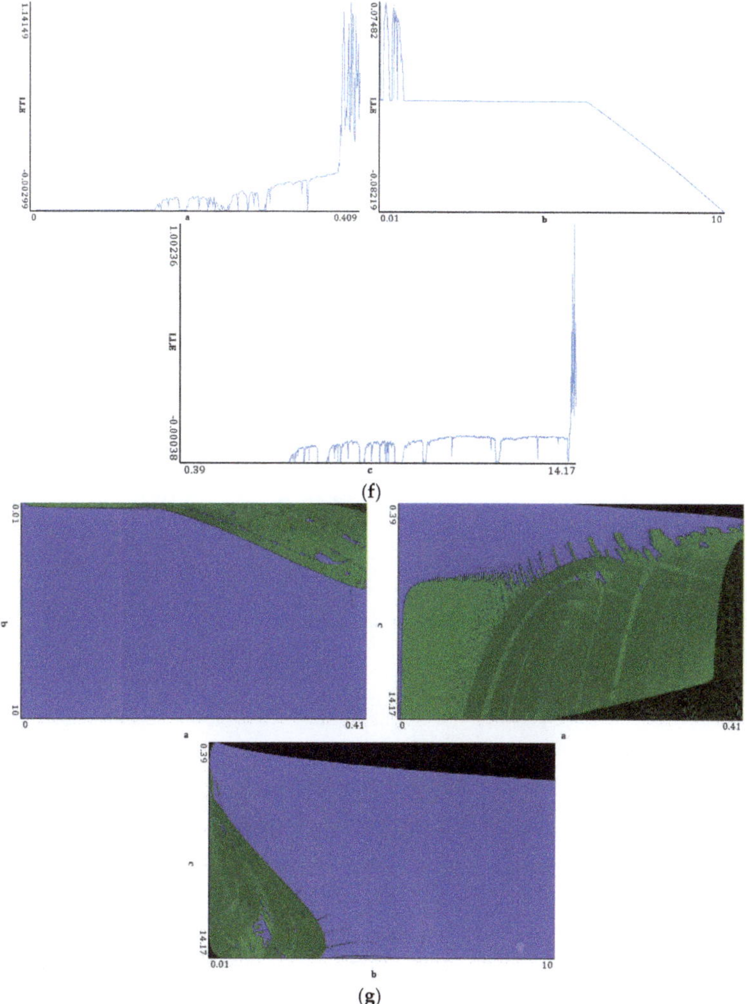

Figure 8. Signal characteristics: (**a**) Time history; (**b**) Time window; (**c**) Chaotic attractor; (**d**) Fourier frequency spectrum; (**e**) Wavelet spectrum; (**f**) Dependence of LLE on the control parameter; (**g**) Lyapunov exponents plane (Rössler attractor).

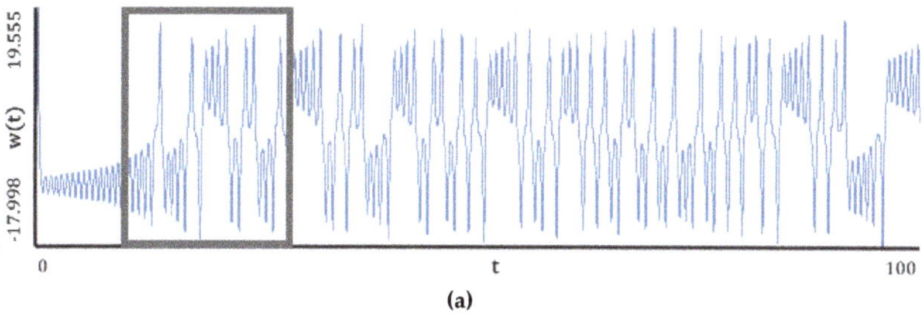

(a)

Figure 9. *Cont.*

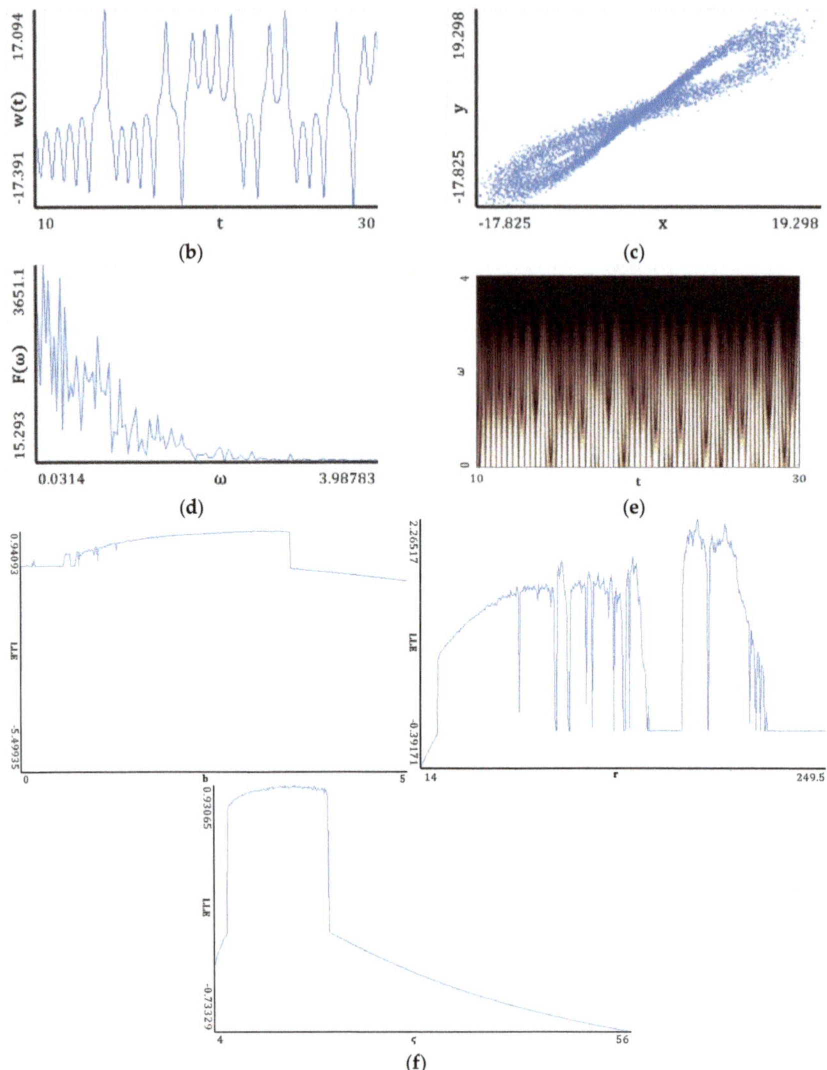

Figure 9. *Cont.*

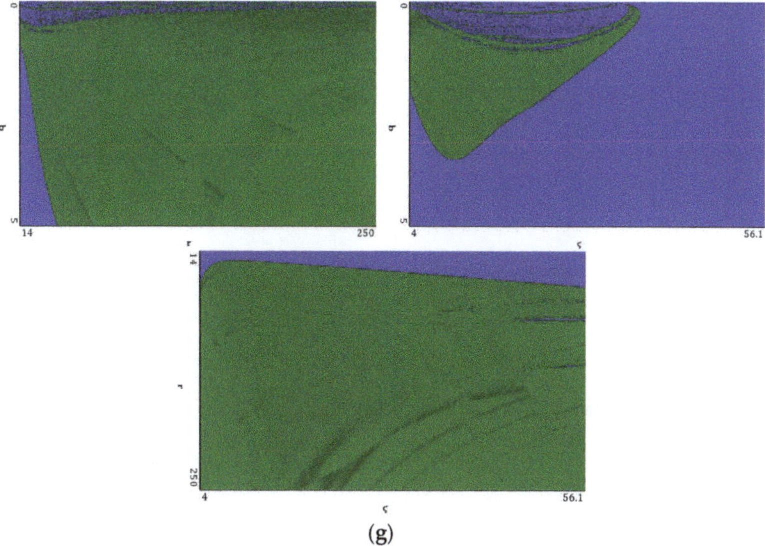

(g)

Figure 9. Signal characteristics: (**a**) Time history; (**b**) Time window; (**c**) Chaotic attractor; (**d**) Fourier frequency spectrum; (**e**) Wavelet spectrum; (**f**) Dependence of LLE on the control parameter; (**g**) Lyapunov exponents plane (Lorenz attractor).

5.1. Logistic Map

A logistic map describes how the population changes with respect to time:

$$X_{n+1} = RX_n(1 - X_n) \tag{27}$$

Here, X_n takes the values from 0 to 1 and presents the population in the n-th year, whereas X_0 denotes the initial population (in the year 0); R is a positive parameter characterizing an increase in the population (computations were carried out for $R = 4$). The first Lyapunov exponent and the Kaplan–Yorke dimension have been estimated by Sprott [33,34]. He has obtained: $\lambda_1 = 0.693147181$, and the Kaplan–Yorke dimension: 1.0.

Figures 5–9 report the following results: (a) signal; (b) signal window; (c) chaotic attractor; (d) Fourier power spectrum; (e) Gauss wavelet of the 8th order, described in Section 4; (f) LLE change depending on the system control parameter; (g) LEs on the control parameters plane (where: ■—only negative Lyapunov exponents, ■—one positive exponent, ■—two positive exponents, ■—three positive exponents).

The power spectrum is noisy and it is not possible to distinguish the dominating frequency. A similar situation is exhibited by the Gauss wavelet, where a large set of frequencies is visible. Dynamics of LLE changing increases for $r > 3$.

As can be seen in Table 1, all computational methods were compared with Benettin's original results. Good coincidence was exhibited by the neural network method, the Rosenstein method, the Kantz method, and the method of synchronization. The Wolf method gave decreased/increased value of LLE in comparison to the original value.

Table 1. Spectrum of Lyapunov exponents and LLEs computed by different methods (logistic map).

LE Spectrum			
Benettin Method		**Neural Network**	
(LEs): 0.69315		LEs: 0.69290	
Dimension Kaplan–Yorke (DKY): 1		DKY: 1	
Kolmogorov-Sinai entropy (KSE): 0.69315		EKS: 0.69290	
Phase volume compression (PVC): 0.69315		PVC: 0.69290	
LLE			
Wolf Method	Rosenstein Method	Kantz Method	Method of Synchronization
LLE: 0.99683	LLE: 0.690553	LLE: 0.69810	LLE: 0.696

5.2. Hénon Map

The Hénon map takes a point (X_n, Y_n) and maps it into another point by the following formulas:

$$
\begin{aligned}
X_{n+1} &= 1 - aX_n^2 + Y_n, \\
Y_{n+1} &= bX_n.
\end{aligned}
\tag{28}
$$

The following parameters are fixed for numerical experiments: $a = 1.4$, $b = 0.3$. Since the Equations (28) do not correspond to a real object, the parameters are replaced with fixed values. Sprott [34] has computed the Lyapunov spectrum and the Kaplan–Yorke dimension of the map using the Benettin method [27] by solving (28). He has obtained the following LEs: $\lambda_1 = 0.419217$, $\lambda_2 = -1.623190$, and the Kaplan–Yorke dimension: 1.258267.

Similarly to the logistic map, the power spectrum exhibits a uniform noisy shape. However, one can distinguish a dominating frequency ($\omega_1 \approx 0,45$). This frequency is also visible on the wavelet spectrum as a region of the largest amplitudes along the whole signal (brighter regions in the graph). Plots of the change in the LLE correlate with bifurcation diagrams for the same interval of changes in the parameters a and b. Dynamics of the LLE changes increases with the increase in both control parameters. Starting with the graphs of LEs for a given set of control parameters, the system mainly remains in a periodic regime, but it exhibits chaotic dynamics for large values of the control parameters.

Beginning from the results shown in Table 2, the majority of the employed computational methods yielded good results. However, the most accurate results were obtained by the neural network method (for whole spectrum of LEs), the Rosenstein method, the Kantz method, and the method of synchronization (in the case of LLEs). The Wolf method gave decreased estimated values of the LLEs.

Table 2. Lyapunov exponents spectrum and LLEs computed by different methods (Hénon map).

Spectrum of LLEs			
Benettin Method		**Neural Network**	
LEs: 0.41919; −1.62316		LEs: 0.41919; −1.62316	
DKY: 1.25826		DKY: 1.25826	
EKS: 0.41919		EKS: 0.41919	
PVC: −1.20397		PVC: −1.20397	
LLEs			
Wolf Method	Rosenstein Method	Kantz Method	Synchronization Method
LLE: 0.38788	LLE: 0.414218	LLE: 0.41912	LLE: 0.40608

Table 3. Lyapunov exponents spectrum and LLEs computed by different methods (generalized Hénon map).

Spectrum of LEs			
Benettin Method		**Neural Network**	
LEs: 0.27628; 0.25770; −4.04053 DKY: 2.13215 EKS: 0.53397 PVC: −3.50656		LEs: 0.29251; 0.27104; −4.04583 DKY: 2.13929 EKS: 0.56355 PVC: −3.48227	
LLEs			
Wolf Method	Rosenstein Method	Kantz Method	Synchronization Method
LLE: 0.45214	LLE: 0.27930	LLE: 0.26601	0.27250

Table 4. Lyapunov exponents spectrum and LLEs computed by different methods (Rössler attractor).

Spectrum of LEs		
Benettin Method	**Neural Network**	
LE: 0.07135; 0.00000; −5.39420 DKY: 2.01323 KSE: 0.07135 PVC: −5.32285	LE: 0.07593; −0.00060; −0.78178 DKY: 2.09635 EKS: 0.07593 PVC: −0.70646	
LLEs		
Wolf Method	Rosenstein Method	Kantz Method
LLE: 0.05855	LLE: 0.0726	LLE: 0.0774

Table 5. Lyapunov exponents spectrum and LLEs computed by different methods (Lorenz attractor).

Spectrum of LEs		
Benettin Method	**Neural Network Method**	
LE: 0.90557; 0.00000; −14.57214 DKY: 2.06214 EKS: 0.90557 PVC: −13.66656	LE: 0.9490; 0.0610; −13.9101 DKY: 2.07261 EKS: 1.0101 PVC: −12.9000	
LLEs		
Wolf Method	Rosenstein Methhod	Kantz Method
LLE: 0.81704	LLE: 0.836	LLE: 0.807185

Table 6. Fourier power spectra (**a**) and Gauss wavelet spectra (**b**) obtained for $\Delta t = 1$, 2 and the LLEs computed by different methods (logistic map).

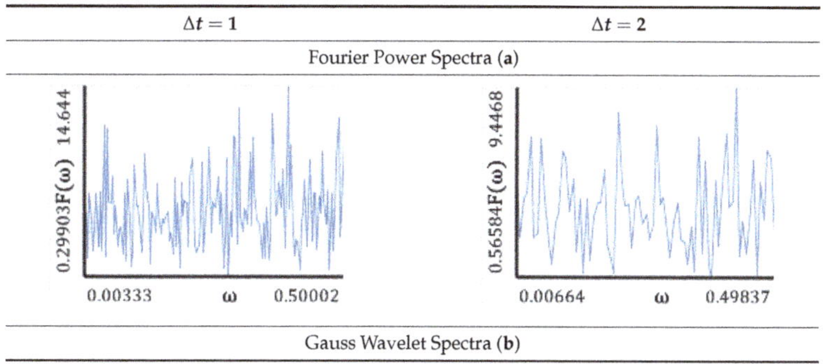

Table 6. *Cont.*

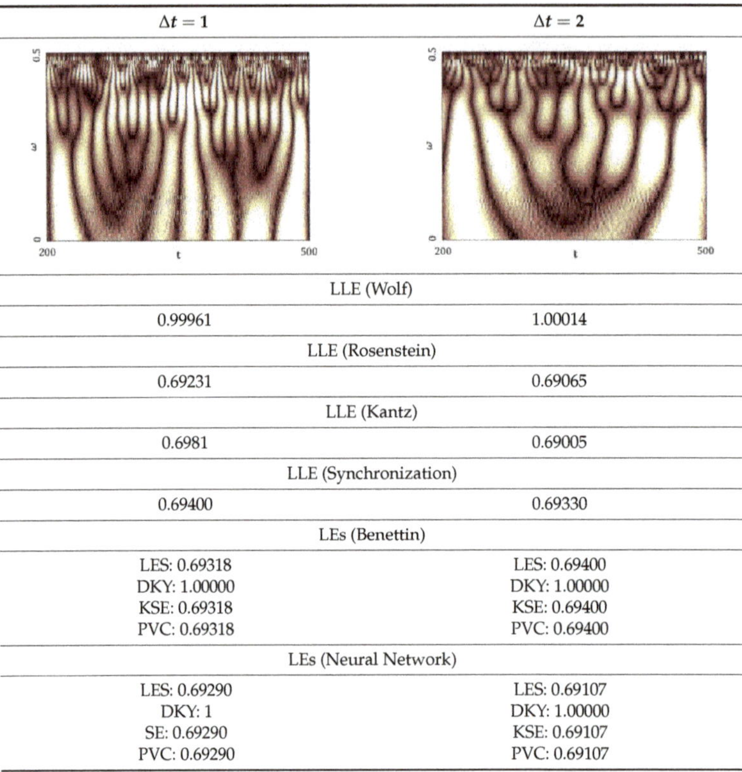

$\Delta t = 1$	$\Delta t = 2$
LLE (Wolf)	
0.99961	1.00014
LLE (Rosenstein)	
0.69231	0.69065
LLE (Kantz)	
0.6981	0.69005
LLE (Synchronization)	
0.69400	0.69330
LEs (Benettin)	
LES: 0.69318 DKY: 1.00000 KSE: 0.69318 PVC: 0.69318	LES: 0.69400 DKY: 1.00000 KSE: 0.69400 PVC: 0.69400
LEs (Neural Network)	
LES: 0.69290 DKY: 1 SE: 0.69290 PVC: 0.69290	LES: 0.69107 DKY: 1.00000 KSE: 0.69107 PVC: 0.69107

Table 7. Fourier power spectra (**a**) and Gauss wavelet spectra (**b**) obtained for $\Delta t = 1$, 2 and the computed LLEs by different methods (Hénon map).

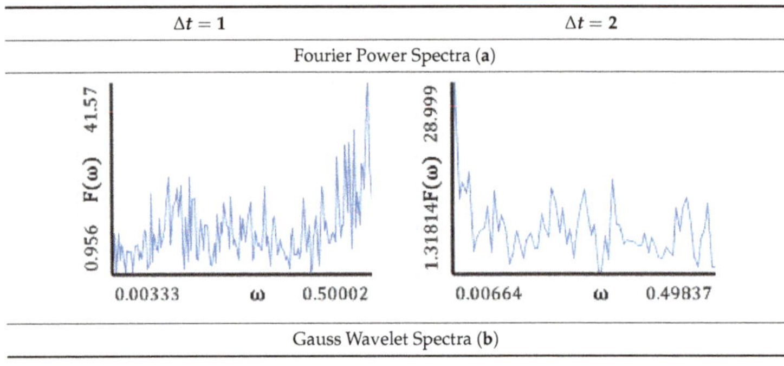

Table 7. *Cont.*

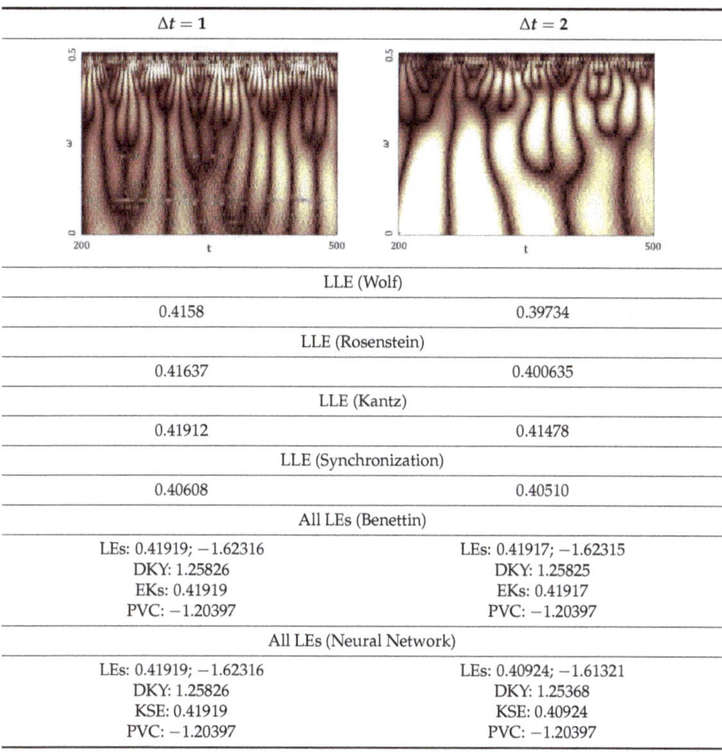

$\Delta t = 1$	$\Delta t = 2$
LLE (Wolf)	
0.4158	0.39734
LLE (Rosenstein)	
0.41637	0.400635
LLE (Kantz)	
0.41912	0.41478
LLE (Synchronization)	
0.40608	0.40510
All LEs (Benettin)	
LEs: 0.41919; −1.62316	LEs: 0.41917; −1.62315
DKY: 1.25826	DKY: 1.25825
EKs: 0.41919	EKs: 0.41917
PVC: −1.20397	PVC: −1.20397
All LEs (Neural Network)	
LEs: 0.41919; −1.62316	LEs: 0.40924; −1.61321
DKY: 1.25826	DKY: 1.25368
KSE: 0.41919	KSE: 0.40924
PVC: −1.20397	PVC: −1.20397

Table 8. Fourier power spectra (**a**) and Gauss wavelet spectra (**b**) obtained for $\Delta t = 1$, 2 and the computed LLEs by different methods (generalized Hénon map).

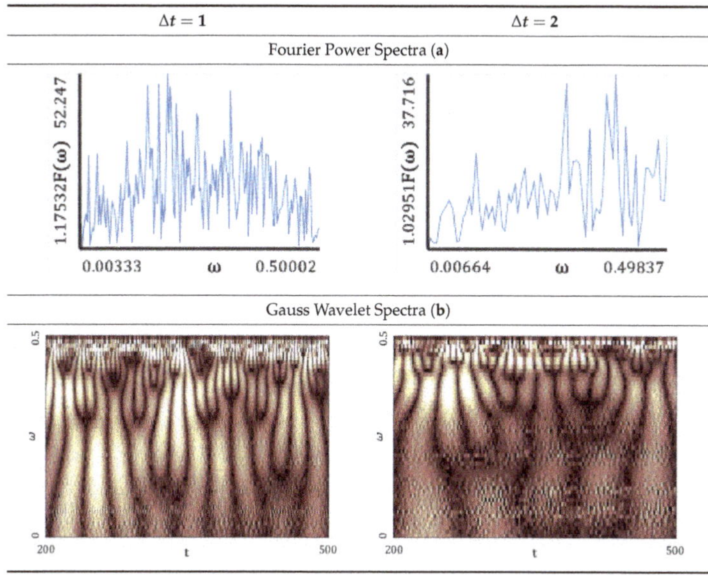

Table 8. *Cont.*

$\Delta t = 1$	$\Delta t = 2$
LLE (Wolf)	
0.45214	0.46706
LLE (Rosenstein)	
0.27930	0.27459 (0.62515)
LLE (Kantz)	
0.26601	0.3359
LLE (Synchronization)	
0.27250	0.27200
All LEs (Benettin)	
LEs: 0.27628; 0.25770; −4.04053 DKY: 2.13215 KSE: 0.53397 PVC: −3.50656	LEs: 0.27487; 0.25631; −4.03774 DKY: 2.13155 EKS: 0.53118 PVC: −3.50656
All LEs (Neural Network)	
LEs: 0.29251; 0.27104; −4.04583 DKY: 2.13929 KSE: 0.56355 PVC: −3.48227	LEs: 0.26304; 0.24387; −4.14321 DKY: 2.12235 KSE: 0.50691 PVC: −3.63630

Table 9. Fourier power spectra and Gauss wavelet spectra obtained for $\Delta t = 0.05$, 0.1, 0.15, 0.2 and the computed LLEs by different methods (Rössler attractor).

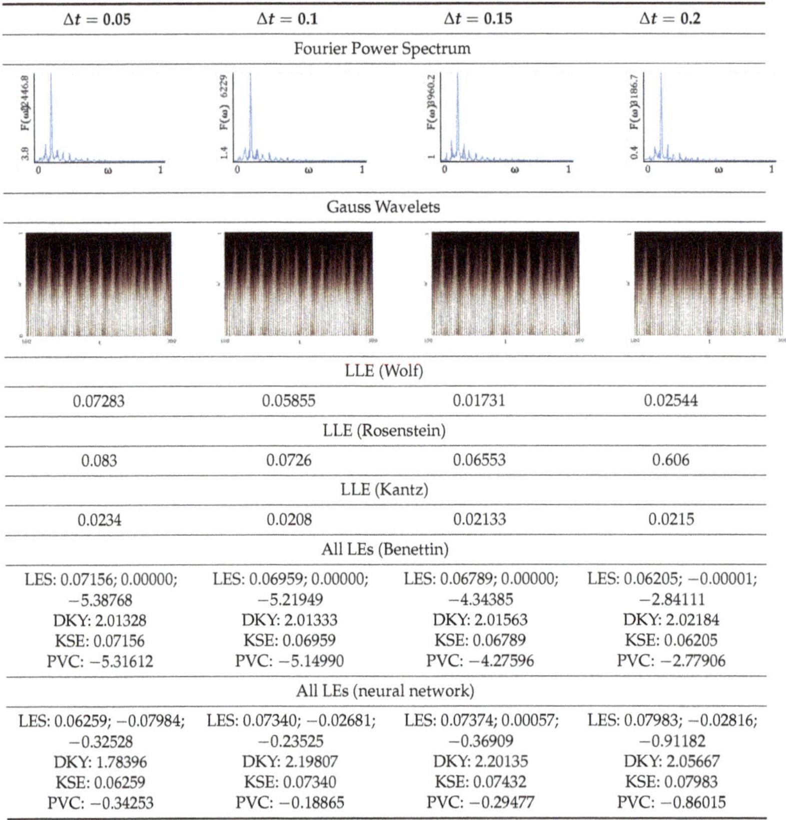

$\Delta t = 0.05$	$\Delta t = 0.1$	$\Delta t = 0.15$	$\Delta t = 0.2$
Fourier Power Spectrum			
Gauss Wavelets			
LLE (Wolf)			
0.07283	0.05855	0.01731	0.02544
LLE (Rosenstein)			
0.083	0.0726	0.06553	0.606
LLE (Kantz)			
0.0234	0.0208	0.02133	0.0215
All LEs (Benettin)			
LES: 0.07156; 0.00000; −5.38768 DKY: 2.01328 KSE: 0.07156 PVC: −5.31612	LES: 0.06959; 0.00000; −5.21949 DKY: 2.01333 KSE: 0.06959 PVC: −5.14990	LES: 0.06789; 0.00000; −4.34385 DKY: 2.01563 KSE: 0.06789 PVC: −4.27596	LES: 0.06205; −0.00001; −2.84111 DKY: 2.02184 KSE: 0.06205 PVC: −2.77906
All LEs (neural network)			
LES: 0.06259; −0.07984; −0.32528 DKY: 1.78396 KSE: 0.06259 PVC: −0.34253	LES: 0.07340; −0.02681; −0.23525 DKY: 2.19807 KSE: 0.07340 PVC: −0.18865	LES: 0.07374; 0.00057; −0.36909 DKY: 2.20135 KSE: 0.07432 PVC: −0.29477	LES: 0.07983; −0.02816; −0.91182 DKY: 2.05667 KSE: 0.07983 PVC: −0.86015

Table 10. Fourier power spectra and Gauss wavelet spectra obtained for $\Delta t = 0.005, 0.01, 0.015, 0.02$ and the computed LLEs by different methods (Lorenz attractor).

$\Delta t = \mathbf{0.005}$	$\Delta t = \mathbf{0.01}$	$\Delta t = \mathbf{0.015}$	$\Delta t = \mathbf{0.02}$
	Fourier Power Spectrum		
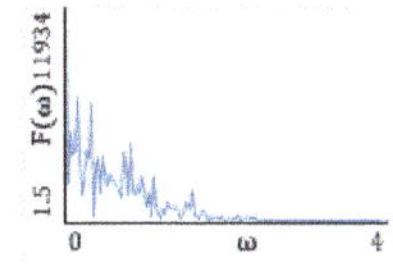	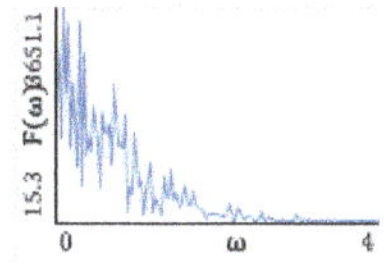	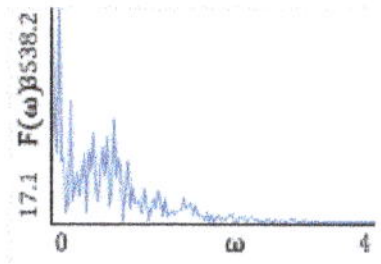	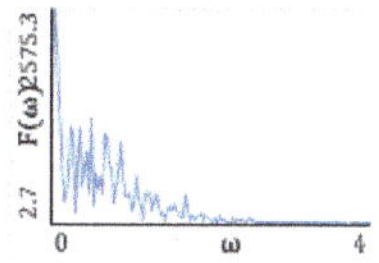
	Gauss Wavelet		
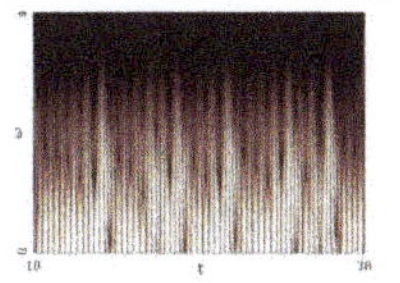	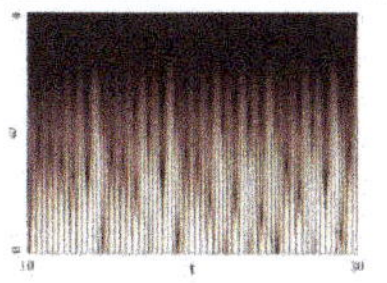	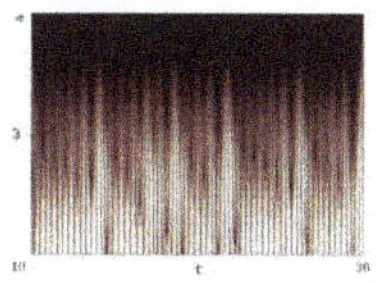	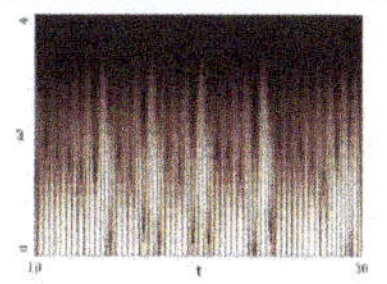
	LLE (Wolf)		
0.9721	0.81704	0.867	0.712
	LLE (Rosenstein)		
0.876	0.836	0.858	0.859
	LLE (Kantz)		
0.898	0.9	0.762667	0.84
	LES (Benettin)		
LES: 0.90632; 0.00000; −14.57297 DKY: 2.06219 KSE: 0.90632 PVC: −13.66666	LES: 0.90523; 0.00000; −14.57179 DKY: 2.06212 KSE: 0.90523 PVC: −13.66656	LES: 0.90551; 0.00000; −14.57163 DKY: 2.06214 KSE: 0.90551 PVC: −13.66613	LES: 0.90596; 0.00000; −14.57086 DKY: 2.06218 KSE: 0.90596 PVC: −13.66490
	LES (Neural Network)		
LES: 0.91677; 0.04404; −6.464 DKY: 2.14864 EKS entropy: 0.96081 PVC: 0.89617	LE: 0.9490; 0.0610; −13.9101 DKY: 2.07261 EKS: 1.0101 PVC: −12.9000	LES: 0.8913; −0.3508; −14.3577 DKY: 2.03765 EKS: 0.8913 PVC: −13.8172	LES: 0.7485; −0.05558; −23.3505 DKY: 2.0296 EKS: 0.7485 PVC: −22.65758

5.3. Hyperchaotic Generalised Hénon Map

To obtain the hyperchaotic Hénon map, one needs to take a point (X_n, Y_n, Z_n) and map it into the following one:

$$\begin{aligned} X_{n+1} &= a - aY_n^2 - bZ_n, \\ Y_{n+1} &= X_n, \\ Z_{n+1} &= Y_n. \end{aligned} \tag{29}$$

The computations were carried out for the following fixed parameters: $a = 3.4$, $b = 0.1$. The Lyapunov spectrum reported in reference [14] is: 0.276; 0.257; 4.040.

One can distinguish a large number of frequencies in the power spectrum. Frequencies with the largest amplitude are located in the interval [0.15; 0.3] (frequencies $\omega_1 - \omega_4$), but the remaining part of the spectrum is noisy. This interval corresponds to the brightest region on the Gauss wavelet, which is correlated with the values of the power spectrum. Changes in LLEs coincide with the bifurcation diagrams constructed for the same intervals of changes in the control parameters a and b. Dynamics of LLEs increases with the increase in the control parameters. As in the case of the Hénon map, the chart of LEs for the selected control parameters exhibits, for a majority of studied parameters, periodic dynamics. It transits into chaos for $a \approx 1.4$, and is almost suddenly shifted into hyperchaos (2 positive LEs).

Good results were obtained by the Benettin, Rosenstein and synchronization methods (divergence from the third decimal place). The neural network yielded slightly increased estimates of two first LEs, whereas the third LE was estimated almost exactly. The Kantz method gave a decreased result in comparison to reference data. The Wolf method resulted in the largest error.

5.4. Rössler Attractor

The following Rössler system of ODEs was investigated:

$$\begin{cases} \dot{x} = -y - z, \\ \dot{y} = x + ay, \\ \dot{z} = b + z(x - c), \end{cases} \tag{30}$$

and the computations were carried out for the following fixed parameters $a = b = 0.2$ and $c = 5.7$.

The original study yielded the Lyapunov spectrum: 0.0714, 0, $-5{:}3943$, and the Kaplan–Yorke dimension equal to 2.0132.

The power spectrum contains the fundamental frequency ω_1, which is accompanied by damped bursts (frequencies $\omega_2 - \omega_{10}$). In the whole time interval, the Gauss wavelet exhibits the brightest region of the fundamental frequency with darker peaks going to zero. Thus, the picture is analogous to the power spectrum. Contrarily to the studied maps, the bifurcation diagrams have a more complex structure. However, there is still correlation with the changes in LLEs for the corresponding control parameters. The parameter b has the smallest influence on the change in LLE. Graphs of LLEs also exhibit a more complex structure. Borders of different vibration kinds have complex forms, which illustrates the increase in the system complexity. Aside from the chaos and hyperchaos zones, there are drops indicating 3 positive LEs. Amabili et al. [35] have suggested to call all chaotic oscillations, for which at least two positive Lyapunov exponents exist, by hyperchaotic.

As far as Table 4 is considered, the best results were yielded by the Benettin and Rosenstein methods. The method of neural networks gave very good results in the case of estimates of two first LEs, but underestimated the third exponent. The Wolf method yielded smaller value of the first exponent compared to the reference data. The most underestimated results were given by the Kantz method.

The carried out numerical experiments showed that changing the sampling frequency did not affect the power spectrum and wavelet spectrum. This was also validated by results obtained by the Benettin, neural networks, and Rosenstein methods, which yielded the results very close to original ones. The Kantz method gave underestimated results for different sampling frequency, correlating with the results obtained for the standard sample size.

5.5. Lorenz Attractor

The system is described by the following ODEs:

$$\begin{cases} \dot{x} = \sigma(y - x), \\ \dot{y} = x(r - z) - y, \\ \dot{z} = xy - bz, \end{cases} \tag{31}$$

where r stands for the normalized Rayleigh number (nondimensional number defining fluid behavior under gradient):

$$r = \frac{g\beta\Delta T L^3}{\nu\chi}. \tag{32}$$

In the above equation, the following notation is used: g—gravity of Earth; L—characteristic dimension of the fluid space; ΔT—temperature difference between fluid walls; ν—kinematic fluid viscosity, χ—thermal conductivity of the fluid; β—coefficient of heat fluid extension; σ—Prandtl number (takes into account heat source property) governed by the following equation

$$\sigma = \frac{\nu}{\alpha} = \frac{\eta C_p}{\aleph},$$ (33)

where: $\nu = \eta/\rho$—kinematic viscosity, η—dynamic viscosity, ρ—density, $\alpha = \frac{\aleph}{\rho C_p}$—temperature transfer coefficient, $\aleph$—heat transfer coefficient, C_p—specific heat capacity under constant pressure; and ρ—information about the geometry of the convective cell.

The following parameters were fixed: $\sigma = 10.0$, $r = 28.0$, $b = 8/3$. The original results are: LEs: 0.9056, 0, −14.5723; the Kaplan–Yorke dimension: 2.06215.

The power spectrum of the attractor decreases uniformly when approaching a finite frequency, and there are no frequencies with a strongly dominating amplitude. The latter observation has been also verified by the Gauss wavelet spectrum. The bifurcation diagrams, similar to those for the Rössler system, exhibit a complex structure, but the correlation to the LLEs change is conserved. The richest/lowest dynamics of LLE is obtained for changing parameter r/σ. Based on the reported graphs of LEs, one can conclude that the system dynamics is fully chaotic. There are also narrow windows of hyperchaotic dynamics.

A comparison of the results reported in Table 5 with the original results exhibits an excellent coincidence of the Benettin method (original results) and the neural network method (+4.79%). The Wolf and Rosenstein methods yielded the underestimated results of the LLE value. The worst estimation has been obtained by Kantz method.

Changing the sampling frequency did not change Fourier and wavelet power spectra. This was also validated by the Benettin and Rosenstein methods, which yield the results very close to the original values in spite of the arbitrary choice of the sampling frequency.

6. Concluding Remarks

The analysis of the dynamics of the studied classical system by different methods leads to a conclusion that the most perspective and useful is the modified method of neural networks [9,10]. It gives excellent convergence to the original results and, as the only one (besides of the Benettin method), allows to compute the spectrum of all Lyapunov exponents. In addition, very good results were obtained by the Rosenstein and Kantz methods for all studied systems. However, this method can be used to estimate only the largest Lyapunov exponents.

As far as the convergence is considered, the Wolf method yielded either over- or underestimated values of LEs. The method of synchronization worked reasonably well for the maps, but it was not useful in studying differential equations (the Rössler or Lorenz systems). The mentioned systems require the use of another type of coupling, which is a drawback of the method.

It should be emphasized that this part of the paper serves as a preliminary study of a more complicated nonlinear continuous structural system, which is studied in Part 2. The carried out analysis of the works devoted to feasible methods for computation of Lyapunov exponents shows that there is no universal, verified, and general method to compute the exact (in the sense of numerics) values of the Lyapunov exponents. This observation leads to the conclusion that there is a need to employ qualitatively different methods to check the reliability of "true chaotic results". Furthermore, the carried out analysis is a helping tool to study systems of an infinite dimension. Such an analysis is the subject of the second part of the paper.

Acknowledgments: This work has been supported by the grants the Russian Science Foundation, RSF 16-19-10290.

Author Contributions: Nikolay P. Erofeev, Vitalyj Dobriyan and Marina A. Barulina performed the numerical study; Anton V. Krysko. and Vadim A. Krysko analyzed the obtained results; Jan Awrejcewicz wrote the paper.

Conflicts of Interest: The authors declare no conflict of interest.

References

1. Wolf, A.; Swift, J.B.; Swinney, H.L.; Vastano, J.A. Determining Lyapunov exponents from a time series. *Phys. D Nonlinear Phenom.* **1985**, *16*, 285–317. [CrossRef]
2. Oseledec, V.I. A multiplicative ergodic theorem. *Trans. Mosc. Math. Soc.* **1968**, *19*, 197–231.
3. Pesin, Y.B. Characteristic Lyapunov exponents and ergodic properties of smooth dynamical systems with invariant measure. *Dokl. Akad. Nauk SSSR* **1976**, *226*, 774–777.
4. Rosenstein, M.T.; Collins, J.J.; de Luca, C.J. A practical method for calculating largest Lyapunov exponents from small data sets. *Phys. D Nonlinear Phenom.* **1993**, *65*, 117–134. [CrossRef]
5. Kantz, H. A robust method to estimate the maximal Lyapunov exponent of a time series. *Phys. Lett. A* **1994**, *185*, 77–87. [CrossRef]
6. Awrejcewicz, J.; Krysko, V.A.; Papkova, I.V.; Krysko, A.V. *Deterministic Chaos in One-Dimentional Continuous Systems*; World Scientific: Singapore, 2016.
7. Stefański, A. Estimation of the largest Lyapunov exponent in systems with impacts. *Chaos Solition Fractals* **2000**, *11*, 2443–2451. [CrossRef]
8. Stefański, A.; Kapitaniak, T. Estimation of the dominant Lyapunov exponent of non-smooth systems on the basis of maps synchronization. *Chaos Solition Fractals* **2003**, *15*, 233–244. [CrossRef]
9. Dobriyan, V.; Awrejcewicz, J.; Krysko, V.A.; Papkova, I.V.; Krysko, V.A. On the Lyapunov exponents computation of coupled non-linear Euler-Bernoulli beams. In Proceedings of the Fourteenth International Conference on Civil, Structural and Environmental Engineering Computing, Cagliari, Italy, 3–6 September 2013; Civil-Comp Press: Stirlingshire, UK, 2013; p. 53.
10. Krysko, A.V.; Awrejcewicz, J.; Kutepov, I.E.; Zagniboroda, N.A.; Dobriyan, V.; Krysko, V.A. Chaotic dynamics of flexible Euler-Bernoulli beams. *Chaos* **2014**, *34*, 043130.
11. Vallejo, J.C.; Sanjuan, M.A.F. Predictability of orbits in coupled systems through finite-time Lyapunov exponents. *New J. Phys.* **2013**, *15*, 113064. [CrossRef]
12. Vallejo, J.C.; Sanjuan, M.A.F. *Predictability of Chaotic Dynamics. A Finite-Time Lyapunov Exponents Approach*; Springer: Berlin, Germany, 2017.
13. Hénon, M. A two-dimensional mapping with a strange attractor. *Commun. Math. Phys.* **1976**, *50*, 69–77. [CrossRef]
14. Baier, G.; Klein, M. Maximum hyperchaos in generalized Henon maps. *Phys. Lett. A* **1990**, *151*, 281–284. [CrossRef]
15. May, R. Simple mathematical model with very complicated dynamics. *Nature* **1976**, *261*, 45–67. [CrossRef]
16. Peitgen, H.-O.; Jürgens, H.; Saupe, D. The Rössler Attractor. In *Chaos and Fractals: New Frontiers of Science*; Springer: Berlin, Germany, 2004; pp. 636–646.
17. Lorenz, E.N. Deterministic nonperiodic flow. *J. Atmos. Sci.* **1963**, *20*, 130–141. [CrossRef]
18. Astafeva, N.M. Wavelet-analysis: Basic theory and examples of applications. *Adv. Phys. Sci.* **1996**, *166*, 1145–1170.
19. Awrejcewicz, J.; Kudra, G.; Wasilewski, G. Experimental and numerical investigation of chaotic regions in the triple physical pendulum. *Nonlinear Dyn.* **2007**, *50*, 755–766. [CrossRef]
20. Dmitriev, A.S.; Kislov, V. *Stochastic Oscillations in Radiophysics and Electronics*; Nauka: Moscow, Russia, 1989.
21. Awrejcewicz, J.; Krysko, A.V.; Papkova, I.V. Routes to chaos in continuous mechanical systems. Part 3: The Lyapunov exponents, hyper, hyper-hyper and spatial–temporal chaos. *Chaos Solition Fractals* **2012**, *45*, 721–736. [CrossRef]
22. Devaney, R.L. *An Introduction to Chaotic Dynamical Systems*; Addison-Wesley: Reading, MA, USA, 1989.
23. Banks, J.; Brooks, J.; Davis, G.; Stacey, P. On Devaney's definition of chaos. *Am. Math. Mon.* **1992**, *99*, 332–334. [CrossRef]
24. Knudsen, C. Chaos without periodicity. *Am. Math. Mon.* **1994**, *101*, 563–565. [CrossRef]
25. Gulick, D. *Encounters with Chaos*; McGraw-Hill: New York, NY, USA, 1992.
26. Awrejcewicz, J.; Krylova, E.U.; Papkova, I.V.; Krysko, V.A. Wavelet-based analysis of the regular and chaotic dynamics of rectangular flexible plates subjected to shear-harmonic loading. *Shock Vib.* **2012**, *19*, 979–994. [CrossRef]
27. Benettin, G.; Galgani, L.; Strelcyn, J.M. Kolmogorov entropy and numerical experiments. *Phys. Rev. A* **1976**, *14*, 2338–2345. [CrossRef]

28. Mackey, M.C.; Glass, L. Oscillation and chaos in physiological control systems. *Science* **1977**, *197*, 287–289. [CrossRef] [PubMed]
29. Hudson, J.L.; Mankin, J.C. Chaos in the Belousov-Zhabotinskii reaction. *J. Chem. Phys.* **1981**, *74*, 6171. [CrossRef]
30. Taylor, G.I. Stability of a viscous liquid contained between two rotating cylinders. *Philos. Trans. R. Soc. Lond. A* **1923**, *223*, 289–343. [CrossRef]
31. Sato, S.; Sano, M.; Sawada, Y. Practical methods of measuring the generalized dimension and the largest Lyapunov exponent in high dimensional chaotic systems. *Prog. Theor. Phys.* **1987**, *77*, 1–7. [CrossRef]
32. Eckmann, J.-P.; Ruelle, D. Ergodic theory of chaos and strange attractors. *Rev. Mod. Phys.* **1985**, *57*, 617–656. [CrossRef]
33. Sprott, J.C. *Chaos and Time Series Analysis*; Oxford University Press: Oxford, UK, 2003.
34. Sprott, J.C. *Elegant Chaos. Algebraically Simple Chaotic Flows*; World Scientific: Singapore, 2010.
35. Amabili, M.; Sarkar, A.; Païdoussis, M.P. Chaotic vibrations of circular cylindrical shells: Galerkin versus reduced-order models via the proper orthogonal decomposition method. *J. Sound Vib.* **2006**, *290*, 736–762. [CrossRef]

Article

Quantifying Chaos by Various Computational Methods. Part 2: Vibrations of the Bernoulli–Euler Beam Subjected to Periodic and Colored Noise

Jan Awrejcewicz [1,*], Anton V. Krysko [2,3], Nikolay P. Erofeev [4], Vitalyi Dobriyan [4], Marina A. Barulina [5] and Vadim A. Krysko [4]

1   Department of Automation, Biomechanics and Mechatronics, Lodz University of Technology, 1/15 Stefanowski St., 90-924 Lodz, Poland
2   Cybernetic Institute, National Research Tomsk Polytechnic University, 30 Lenin Avenue, 634050 Tomsk, Russia; anton.krysko@gmail.com
3   Department of Applied Mathematics and Systems Analysis, Saratov State Technical University, 77 Politechnicheskaya, 410054 Saratov, Russia
4   Department of Mathematics and Modeling, Saratov State Technical University, 77 Politechnicheskaya, 410054 Saratov, Russia; erofeevnp@mail.ru (N.P.E.); dobriy88@yandex.ru (V.D.); tak@san.ru (V.A.K.)
5   Precision Mechanics and Control Institute, Russian Academy of Science, 24 Rabochaya Str., 410028 Saratov, Russia; marina@barulina.ru
*   Correspondence: jan.awrejcewicz@p.lodz.pl; Tel.: +48-42-6312-225

Received: 17 January 2018; Accepted: 1 March 2018; Published: 5 March 2018

Abstract: In this part of the paper, the theory of nonlinear dynamics of flexible Euler–Bernoulli beams (the kinematic model of the first-order approximation) under transverse harmonic load and colored noise has been proposed. It has been shown that the introduced concept of phase transition allows for further generalization of the problem. The concept has been extended to a so-called noise-induced transition, which is a novel transition type exhibited by nonequilibrium systems embedded in a stochastic fluctuated medium, the properties of which depend on time and are influenced by external noise. Colored noise excitation of a structural system treated as a system with an infinite number of degrees of freedom has been studied.

Keywords: geometric nonlinearity; Bernoulli–Euler beam; colored noise; noise induced transitions; true chaos; Lyapunov exponents; wavelets

---

1. Introduction

There are some well-recognized and thoroughly investigated scenarios of transition from regular to chaotic vibration of deterministic systems, such as the well-known Ruelle–Takens scenario, in which the key role is played by only three frequencies [1]. It was found that the "noisy" behavior was exhibited by a "strange attractor" occurred after three successive limit cycle bifurcations. In addition, a period doubling bifurcation route to chaos as well as a scenario of transition by means of intermittency were also detected and studied [2]. However, a stochastic character of the matter may induce more interesting regimes of nonlinear vibrations than that reported in the available literature [3]. Transitions from one nonlinear state to another are analogous to equilibrium phase transitions and they are met in nonequilibrium processes under deterministic external inputs. A question arises: Is it possible to extend the classical transitions to processes in which a key role is played by noise? It implies theoretical investigations of the transitions induced by noise based on the numerical study. The investigation of deterministic chaotic multi-dimensional systems has been carried out (e.g., [4–9]). The aim of the present work was to extend the mentioned approaches to the case of the noise-induced transitions into chaos.

The investigation of complex vibrations of continuous systems belongs to challenging problems of modern dynamics. In the works of Amabili [10,11], distributed mechanical systems have been considered in the form of a shallow shell of double curvature and a closed cylindrical shell with 16 degrees of freedom. The problem has been solved by Galerkin's method, but the number of similar works is limited. In most of the available papers, reduced-model approaches have been used to simplify these considerations. Namely, one can find numerous papers attempting to significantly simplify the studied problem of an infinite dimension to one- or two-degree-of-freedom reduced models. The obtained results are usually validated under strong assumptions and can be treated rather as qualitative since an increase in the number of modes often yields essentially different outcome.

One of the most important problems encountered when studying chaotic vibrations is the reliability of the obtained results in the case of true/real chaos. True/real chaos means a reliable and validated chaotic solution obtained numerically, and is contrary to a possible "false chaotic solution" caused by the errors associated with the used numerical algorithms as well as time and spatial steps. A more detailed description can be found in Reference [12]. The majority of investigations rely on numerical simulations, which may accumulate the errors yielding false results regarding the estimation of true chaos. Therefore, one of the aims of the present paper was to validate the truth of chaotic orbits by different numerical methods, since the obtained results may depend on the employed numerical method.

Proper modeling and the way of solving the spatiotemporal nonlinear dynamics of Euler–Bernoulli beams and other structural members governed by nonlinear PDEs plays a crucial role in obtaining true/reliable results. The majority of the research devoted to dynamics of continuous mechanical systems has its origin in engineering and, depending on the needs, the governing PDEs should satisfy different requirements regarding the results accuracy and duration of the computations. For instance, in static problems, it is rather expected to get highly accurate models, which usually require long computational time since the commonly used program packages are based on the FEM (finite element method) and/or the FDM (finite difference method) as well as the higher-order Galerkin methods, the Ritz-type methods, etc. Contrarily, dynamic problems (often) require control, which means from the point of view of complexity of the governing PDEs that the developed model should be rather dedicated to the considered task and the nonlinear PDEs should be appropriately reduced to get reliable and relatively fast numerical results. In general, there is no way to solve the infinite-dimension nonlinear problems analytically.

Thus, the knowledge and experience about proper modeling of nonlinear vibrations of structural members as well as the way of getting solution to the governing PDEs with the supplemented boundary and initial conditions are the key factors in achieving reliable results.

To exhibit the importance of the mentioned problem, we used the classical Lorentz [13] results, showing that, depending on the time step, the used numerical schemes (Euler and Runge–Kutta) and Taylor series truncation, one can get the so-called "computational chaos", i.e., a spurious solution not validated by either real laboratory experiment or qualitatively different numerical approaches. As reported by Yee and Sweby [14], even small time steps and the employment of the classical Euler, Runge–Kutta, predictor-corrector schemes may yield spurious solutions (sometimes coexisting with true solutions). It may also happen that the decrease in the time step may yield worse computational results. For instance, Corless et al. [15] pointed out that a true/correct chaotic orbit can be even suppressed.

Note that the investigation of spatial dynamics of structures involves distribution in time and space and, usually, the problem should be reduced to that of many nonlinear ODEs and possibly AEs (algebraic equations). However, the first-mode approximations are usually used to investigate chaotic vibrations of simple 1D structural members (beams). As mentioned above, this approach requires to be rather strictly realized in real/engineering problems (see, e.g., [16–19]).

In the following, the description of numerous works devoted to study nonlinear beams, including the Euler–Bernoulli beams, is omitted and we refer only to the recent tendency in analyzing nonlinear dynamics of beams taking into account a stochastic/random excitation.

Bar-Yoseph et al. [20] employed the space-time spectral element method for a nonlinear Euler–Bernoulli beam using a generalized Galerkin approach as well as a discontinuous mixed Galerkin method for the temporal discretization. However, they considered only the two first modes.

Based on short datadets, Dunne and Ghanbari [21] predicted extreme exceedance probabilities matched with experimental investigations of highly nonlinear clamped-clamped beam vibrations driven by band-limited white noise.

Ge and Xu [22] proposed a stochastic nonlinear dynamical model of vibrations of a flexible beam under axial excitation, considering random environmental factors. The governing PDEs were simplified using the stochastic average theory, and then the methods of Lyapunov exponents and boundary classification with the account of diffusion process were employed to study the stochastic stability of the trivial solutions. Then, the stochastic Hopf bifurcation was studied. However, the problem was limited to investigating only one second-order ODE.

Cacciola et al. [23] analyzed the vibrational response of a beam with an edge nonpropagating crack by means of a stochastic analysis to detect the presence and the location of structural damage. Numerical investigation was used to show that a cantilever beam exhibited high sensitivity of the skewness coefficient of the rotational degrees of freedom. However, only third- and fourth-order statistical quantities of the response were evaluated, and the obtained results were not validated experimentally.

Sahoo et al. [24] studied nonlinear transverse vibration of a simply-supported traveling Euler–Bernoulli beam under principal parametric resonance and in presence of internal resonance. However, the problem was essentially limited to studying only the principal parametric resonance of the first mode by employing the method of multiple scales.

Lan and Qin [25] investigated the energy harvesting from the horizontal coherent resonance of a vertical cantilever beam subjected to the vertical base excitation. They demonstrated the horizontal coherence resonance when the beam was excited by a vertical white noise. However, derivation of the governing PDEs was not shown, but only the experimental results were reported.

Xu and Ma [26] investigated, in a rigorous mathematical way, the existence of a compact random attractor for the floating beam equation with strong damping and white noise. However, their results seem to be mathematically-oriented. The authors did not present any numerical results, and hence direct application of the obtained results to real problems is rather limited.

It is known and commonly accepted that the main characteristic feature of deterministic chaos is its essential dependence on the initial conditions. The definition of chaos, initially formulated by Devaney in 1989 [27], consists of three parts. In addition to the essential dependence on the initial conditions, there is a chaotic mixing condition referred to as the transitivity and regularity condition, measured by a density of periodic points. On the other hand, in 1992, Banks et al. [28] proved that the transitivity and regularity conditions imply the sensitivity to the initial conditions.

A more rigorous chaos definition, given by Knudsen [29] also exists, where a function defined on a bounded metric space is chaotic if it has a dense orbit and exhibits essential dependence on the initial conditions.

Owing to chaos definition proposed by Gulick [12], chaos exists if either there is an essential dependence on the initial conditions or a chaotic function has a positive Lyapunov exponent in its each point, and hence if it finally does not tend to a periodic or quasi-periodic orbit.

In this paper, similar to in Part 1, we employed chaos definition given by Gulick [12], i.e., we validated true chaotic solutions by employing computations of the Lyapunov exponents. The obtained solutions depend on the chosen kinematic hypothesis, boundary and initial conditions, number of beam partitions in the case of the FDM (finite difference method), the chosen method of solutions to the Cauchy problems, and the chosen time step.

To detect true chaotic vibrations when numerically solving the problem devoted to nonlinear vibrations of a package consisting of two beams with a clearance and subjected to transversal sinusoidal load and the colored noise, the following complex investigations were employed:

1. Since PDEs are reduced to ODEs (Cauchy problem) using the FDM of second-order accuracy, their solution essentially depends on the number of beam length partitions (nodes). We need to find a number of partitions n, for which the solution coincided with the case of using $n + 1$ partitions. Furthermore, we want to achieve convergence even with respect to time history in the case of chaotic orbits. In monograph [30] dealing with a similar problem, the convergence was achieved only with respect to the periodic vibrations, and in the case of chaotic vibrations, only integral convergence was accepted, i.e., coincidence of power spectra was used to validate true chaos.

2. The Cauchy problem was solved numerically and it is known that it depends on the chosen method and the integration step, which is why we chose different methods to validate the computational results: fourth order the Runge–Kutta method (RK4) and the second order Runge–Kutta method (RK2) [31], the fourth order Runge–Kutta–Fehlberg method (RKF45) [32,33], the fourth order Cash–Karp method (RKCK) [34], the eighth order Runge–Kutta–Prince–Dormand method (RK8PD) [35], the implicit second order Runge–Kutta method (rk2imp), and the fourth order Runge–Kutta implicit method (rk4imp). The implicit method makes it possible to include an arbitrary form of the used matrix of the related coefficients, whereas the RKF45, RKCK, and RK8PD methods allow for the automatic change of the computational step as well as to control errors introduced by integration.

3. For each of the introduced numbers of partitions of the beam length and the solutions to the Cauchy problems, the time histories (vibration signals), 2D and 3D phase portraits, the Fourier power spectra, the Morlet wavelets, snapshots of beam deflections, and Poincaré maps were constructed. For the chosen signal, a 2D wavelet spectrum was also constructed. The following mother wavelets were employed: Haar [36]; Shannon–Kotelnikov and Meyer [37]; Daubechies wavelets from db2 up to db16 [38]; coiflets, simlets, and the Morlet and complex Morlet wavelets [39]; and the wavelets based on the derivative of the Gauss function of the order higher than eight. However, the Haar and Shannon–Kotelnikov wavelets were unsuitable for our purpose. Namely, the first one was badly localized with respect to frequency, and the second one with respect to time.

The analysis of the wavelet spectra yielded by the Daubechies wavelets, coiflets and simlets showed that an increase in the filter order results in an increase in the frequency deflection/localization. The results obtained by the Daubechies wavelets, simlets and coiflets were practically the same, which validates them. However, their frequency localization was not enough suitable to be reliable while studying the character of vibrations of the studied continuous mechanical system.

As expected, the results obtained based on the derivative of the Gauss functions yielded an increase in the frequency resolution with an increase in the derivative order. The power spectra obtained based on the Meyer wavelet [40,41] (a smoothened variant of the Shannon–Kotelnikov wavelet) were better localized in the low-frequency band than the Morlet power spectra. However, the latter ones fit the higher part of the frequency spectrum better. In the following, we give the results based on the mother Morlet wavelets. In Morlet wavelets, there is a lack of scaling function φ, the function ψ does not have a compact carrier and it is given explicitly. The complex Morlet and Gauss wavelets exhibit better localization with respect to frequency, but their real counterpart is better localized in time. Therefore, to study complex vibrations of dynamical continuous systems (beams), one can use both real and complex Morlet wavelets based on the derivatives of the Gauss functions of the order higher than sixteen.

4. Since we employed the Gulick [12] definition of chaos, we needed to compute and validate signs of the largest Lyapunov exponents (LLEs). Spectra of Lyapunov exponents were estimated based

on the Kantz [42], Wolf [43] and Rosenstein [44] methods. The results were eventually accepted if they agreed to the fourth decimal place.

Vibrations of geometrically nonlinear Bernoulli–Euler beam (kinematic model of the first approximation) were studied as a system with many degrees of freedom subjected to colored noise. White noise states for a general stationary probabilistic process $X(t)$ with a constant spectral density. The term "white" corresponds to the white light, the spectrum of which combines the whole palette of colors in the visible spectrum part. The correlation white noise function is governed by the formula $B(t) = \sigma^2 \delta(t)$, where σ^2 is a constant, and $\delta(t)$ stands for the delta function. An essentially nonlinear character of the behavior of plates and shells under loading by noise pressure has been found in several works. The Gaussian white noise model is very useful and appropriate to study different natural processes. The colored noise spectrum corresponds to the distribution of the visible part of the light spectrum.

2. Problem Statement and the Mathematical Model

We considered a one-layer isotropic elastic beam (Figure 1) representing a 2D area of the space R^2 with the attached coordinate system introduced in the following way: in the beam body, a so-called reference line, also called a middle line, $z = 0$ was fixed; the axis OX was directed from the left to the right along the middle line. The beam area Ω, in the given coordinate system, was defined as follows: $\Omega = \{x \in [0; a]; -h \le z \le h\}, 0 \le t \le \infty$. The following notation was employed: $2h$—beam height, a—beam length.

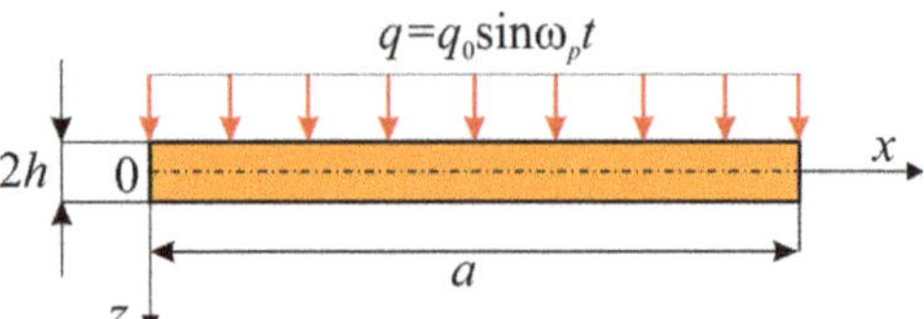

Figure 1. Scheme of the investigated beam.

The mathematical model was constructed based on the Bernoulli–Euler hypothesis considering a nonlinear relation between deformations and displacements in the von Kármán form. The governing PDEs in displacements had the following form [45]:

$$
\begin{aligned}
&E2h\left\{\frac{\partial^2 u}{\partial x^2} + L_3(w, w)\right\} - 2h\frac{\gamma}{g}\frac{\partial^2 u}{\partial t^2} = 0, \\
&E2h\left\{L_1(u, w) + L_2(w, w) - (2h)^2\left(\frac{1}{12}\right)\frac{\partial^4 w}{\partial x^4}\right\} + q + \dot{q} - 2h\frac{\gamma}{g}\frac{\partial^2 w}{\partial t^2} - 2h\varepsilon\frac{\gamma}{g}\frac{\partial w}{\partial t} = 0, \\
&L_1(u, w) = \frac{\partial^2 u}{\partial x^2}\frac{\partial w}{\partial x} + \frac{\partial u}{\partial x}\frac{\partial^2 w}{\partial x^2}, L_2(w, w) = \frac{3}{2}\frac{\partial^2 w}{\partial x^2}\left(\frac{\partial w}{\partial x}\right)^2, L_3(w, w) = \frac{\partial^2 w}{\partial x^2}\frac{\partial w}{\partial x},
\end{aligned}
\tag{1}
$$

where: $w(x, t)$ is the beam deflection, $u(x, t)$ is the displacement along the OX axis, ε is the dissipation coefficient, $q = q(x, t)$ is the transverse load, $\dot{q}(t)$ is the colored noise, E is Young's modulus, γ is the specific volume beam weight, and g is the gravity of Earth.

The following non-dimensional quantities were introduced:

$$
\begin{aligned}
&\overline{w} = \frac{w}{2h}, \quad \overline{u} = \frac{ua}{(2h)^2}, \quad \overline{x} = \frac{x}{a}, \quad \lambda = \frac{a}{2h}, \quad \overline{q} = q\frac{a^4}{(2h)^4 E}, \\
&\overline{t} = \frac{t}{\tau}, \quad \tau = \frac{a}{c}, \quad c = \sqrt{\frac{Eg}{\gamma}}, \quad \overline{\varepsilon} = \varepsilon\frac{a}{c}.
\end{aligned}
\tag{2}
$$

The system in Equation (1), taking into account Equation (2), took the following simple counterpart form

$$\frac{\partial^2 u}{\partial x^2} + L_3(w, w) - \frac{\partial^2 u}{\partial t^2} = 0,$$
$$\frac{1}{\lambda^2}\left\{ L_2(w, w) + L_1(u, w) - \left(\tfrac{1}{12}\right)\frac{\partial^4 w}{\partial x^4} \right\} - \frac{\partial^2 w}{\partial t^2} - \varepsilon\frac{\partial w}{\partial t} + q + \dot{q} = 0, \tag{3}$$

(bars over non-dimensional quantities were omitted).

Equation (3) was supplemented by the following boundary conditions: simple support

$$w(0, t) = w(1, t) = u(0, t) = u(1, t) = \partial^2 w(0, t)/\partial x^2 = \partial^2 w(1, t)/\partial x^2 = 0 \tag{4}$$

or clamping

$$w(0, t) = w(1, t) = u(0, t) = u(1, t) = \partial w(0, t)/\partial x = \partial w(1, t)/\partial x = 0 \tag{5}$$

on both beam ends.

In the case of mixed boundary conditions (clamping and simple support):

$$w(0, t) = w(1, t) = u(0, t) = u(1, t) = \partial w(0, t)/\partial x = \partial^2 w(1, t)/\partial x^2 = 0 \tag{6}$$

The initial conditions were as follows:

$$w(x, t)\big|_{t=0} = \frac{\partial w(x, t)}{\partial t}\bigg|_{t=0} = u(x, t)\big|_{t=0} = \frac{\partial u(x, t)}{\partial t}\bigg|_{t=0} = 0 \tag{7}$$

3. Methods of Solution

3.1. FDM Method

Wes substituted the differential operators with respect to x in Equations (3)–(7) by difference operators for the function $w(x, t)$, $u(x, t)$ with the help of the FDM with approximation $O(h^2)$. PDEs in Equation (3) were reduced to the following ODEs

$$\ddot{w}_i + \varepsilon_1\dot{w}_i = L_{1,h}(w_i, u_i), \quad \ddot{u}_i = L_{2,h}(w_i, u_i), \quad i = 0, \ldots, n, \tag{8}$$

where n is the number of partitions along x, c stands for the spatial step, and the difference operators $L_{1,c}$ and $L_{2,c}$ have the following form:

$$L_{1,c}(w_i, u_i) = \frac{1}{\lambda^2}\left(-\frac{1}{12}\frac{1}{c^4}(w_{i+2} - 4w_{i+1} + 6w_i - 4w_{i-1} + w_{i-2}) + \frac{1}{2c}(w_{i+1} - w_{i-1}) \right.$$
$$\times \frac{1}{s^2}(u_{i+1} - 2u_i + u_{i-1}) + \frac{1}{2c}(w_{i+1} - w_{i-1})\frac{1}{s^2}(u_{i+1} - 2u_i + u_{i-1}) + \left(\frac{1}{2c}(w_{i+1} - w_{i-1})\right)^2\frac{1}{c^2}(w_{i+1} - 2w_i + w_{i-1})+$$
$$\left. \frac{1}{c^2}(w_{i+1} - 2w_i + w_{i-1})\left(\frac{1}{2s}(u_{i+1} - u_{i-1}) + \frac{1}{8s^2}(w_{i+1} - w_{i-1})(w_{i+1} - w_{i-1})\right)\right),$$
$$L_{2,c}(w_i, u_i) = \frac{1}{c^2}(u_{i+1} - 2u_i + u_{i-1}) + \frac{1}{2c}(w_{i+1} - w_{i-1})\frac{1}{c^2}(w_{i+1} - 2w_i + w_{i-1})$$

In Equations (8), it is necessary to consider the values at the boundary points, which are defined by the boundary conditions. For the boundary conditions in Equation (4): $w_{-i} = -w_i$, $w_0 = 0$, $w_n = 0$, $u_0 = 0$, $u_n = 0$ (simple support). For the boundary conditions in Equation (5): $w_{-i} = -w_i$, $w_0 = 0$, $w_n = 0$, $u_0 = 0$, $u_n = 0$ (clamping).

The initial conditions in Equation (7) can be rewritten in the difference form as follows:

$$w(x_i)\big|_{t=0} = 0; \ u(x_i)\big|_{t=0} = 0; \ \dot{w}(x_i)\big|_{t=0} = 0; \ \dot{u}(x_i)\big|_{t=0} = 0, \ (i = 0, \ldots, n).$$

3.2. Lyapunov Exponents

Wolf et al. [43] proposed an algorithm allowing for estimation of positive Lyapunov exponents based on the time series. It was shown that Lyapunov exponents correspond to either exponential fast divergence or convergence of the neighborhood orbits in the studied phase space. The employed concept of computation comes from the earlier method applied only to analytically defined mathematical models. The method traces a long term increase in a small volume chosen in the studied attractor and yields the largest Lyapunov exponent (LLE) based on one time series. The method does not require any knowledge about the evolutionary equations and phase coordinates. The algorithm for computing the LLE was described in Part 1 of our paper. As a reminder, the Kantz [42] and Rosenstein [44] methods are based on constructing the time-dependent function and take into account divergence of the nearby trajectories in the phase space in a certain time period. Then, they try to find the straightest part of this function (which is not always possible). A tangent of the inclination angle of this straight line yields the LLE.

3.3. Case Studies and Results Analysis

Owing to the above-described methodology, we began with investigating the convergence of the FDM with approximations $O(h^2)$ and the Runge–Kutta methods as well as with comparing the results obtained by different Runge–Kutta methods. The comparisons were made by taking $n = 40; 80; 120; 160; 200; 240; 280; 320; 360; 400; 440$ nodes in the FDM when transiting PDEs in Equation (3), the boundary conditions in Equation (5) and the initial conditions in Equation (7) to ODEs (Cauchy problem). The Cauchy problem was solved by the Runge–Kutta methods of the second, fourth and eighth orders (Figure 2b reports the results obtained for $n = 360$ yielded by different kinds of Runge–Kutta methods). The most reliable results were obtained for the eighth order Runge–Kutta method in the Prince–Dormand modified version. Convergence of the vibration periodic and chaotic signals was investigated. Figures 2 and 3 present time histories/signals (Figure 2a), phase portraits (Figure 2b), power spectra (Figure 2c) and the Morlet wavelet spectra (Figure 3) for the periodic vibrations depending on the nodes number $n = 40; 80; 160$. Convergence of FDM was fully achieved for $n = 80$. All the mentioned vibration characteristics fully coincided.

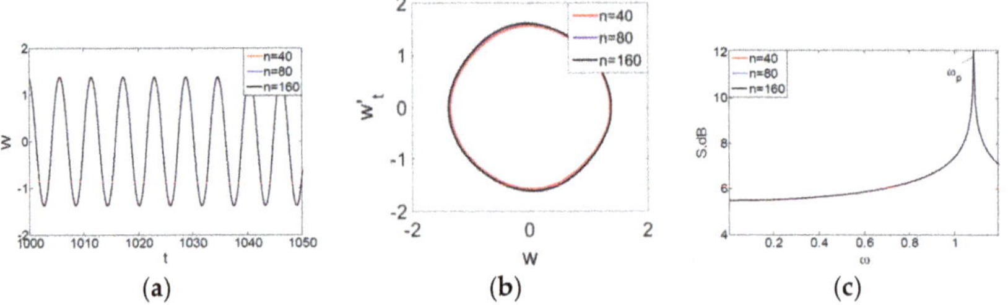

Figure 2. Convergence of the FDM (finite difference method) with respect to the vibrations signal and Fourier power spectrum (periodic orbits). (**a**) Signal ($n = 40; 80; 160$); (**b**) Phase portraits ($n = 40; 80; 160$); (**c**) Fourier power spectrum ($n = 40; 80; 160$).

Then, we studied convergence of the FDM for chaotic vibrations. Figure 4a presents the vibration signals $w(0.5, t)$ obtained for $n = 120; 240; 320; 360; 400$. For $n = 120$, the deflection essentially differed from the deflection obtained for $n = 240; 320; 360; 400$ (for which the deflections coincided).

Owing to Point 3 of the described methodology, power spectra, phase portraits, and Morlet wavelet spectra for $\omega \in [0.2; 5]$ are reported in Table 1. These characteristics were considered for $n = 120; 320; 360; 400$. In the power spectrum, for $n = 120$, the independent frequencies ω_1, ω_3, ω_5, ω_6, ω_8, ω_9, ω_{10}, ω_{11}, ω_{12}, ω_{13}, ω_{15}; the dependent frequencies $\omega_{14} = \frac{2}{5}\omega_p$, $\omega_7 = \frac{3}{5}\omega_p$;

and a chaotic component were detected. An increase in the number of nodes changed the signal frequency spectrum (Fourier power spectrum). Eleven frequencies were detected for $n = 320$ and twelve for $n = 360; 400$. The phase portraits present washed out clouds converging to a ring with the increase in the number of nodes up to $n = 320; 360; 400$. For $n = 120$, the wavelet spectrum exhibits three frequencies ω_p, ω_3 and ω_{14}, and an increase in the number of nodes yielded only two frequencies—ω_p and ω_3. It means that convergence with respect to power spectra, phase portraits and wavelet spectra was achieved.

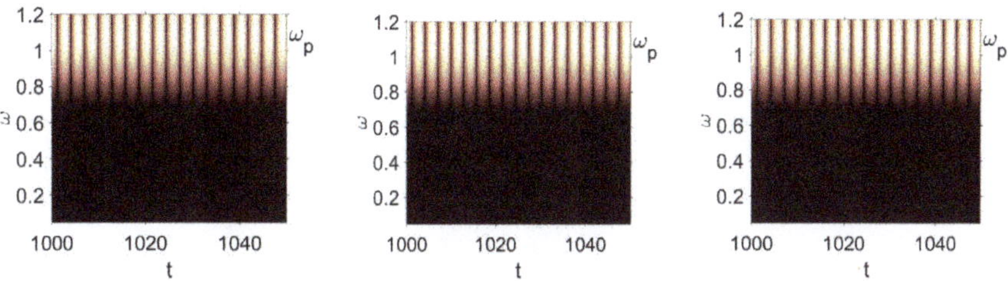

Figure 3. Convergence of the FDM with respect to the Morlet wavelet spectrum (periodic orbits).

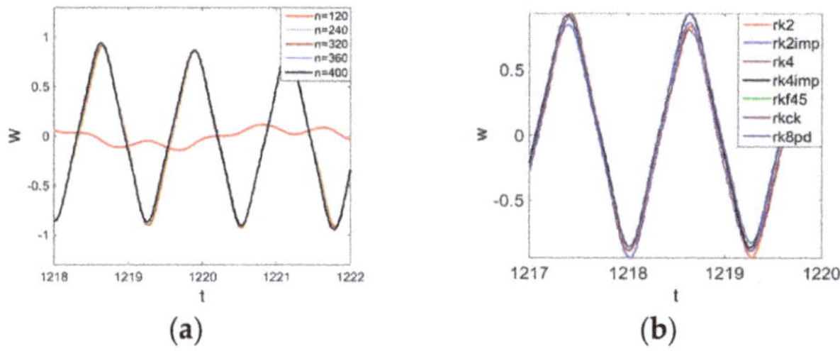

(**a**)   (**b**)

Figure 4. Signals obtained for $n = 120; 240; 320; 360; 400$; (**a**) Convergence of $w(0.5; t)$, $t \in [1218; 1222]$ vs. Number of Nodes; (**b**) Convergence of $w(0.5; t)$, $t \in [1217; 1220]$ vs. Numerical Method Types.

All Lyapunov exponents (Table 2) obtained by three mentioned methods are positive, and each of the methods is convergent with respect to n $(a/(2\,h) = 50)$.

To study behavior of nonlinear beams under harmonic $q = q_0 \sin(\omega_p t)$ load and colored noise of amplitude C_n, a package of programs allowing for construction and analysis of the power spectra and Lyapunov exponents was developed. Beam vibrations were traced on the time interval $t \in [0; 2024]$, the number of partitions in the FDM was $n = \{80; 400\}$, and the ratio of the beam length to the beam thickness was $\lambda = 50$. All results were obtained for the beam center. Vibration signals, Fourier power spectra, Morlet wavelet-spectra, phase and modal portraits, autocorrelation functions and LLEs were analyzed for the given type of colored noise.

When carrying out the numerical study, we chose a frequency such that the addition of low-amplitude noise did not change the system state. This motivated us to consider the noise amplitude $C_n = 5,000$ being compatible with the amplitude of periodic excitation.

The spectrum density of the employed colored noises is proportional to $1/f^{\alpha}$, where f stands for the frequency of the spectrum, and α takes the following values: $\alpha = -2$—purple noise; $\alpha = -1$—blue noise; $\alpha = 0$—white noise; $\alpha = 1$—pink noise; and $\alpha = 2$—brown noise.

The values for all time series for colored noise were found in the interval $[-1; 1]$ and then multiplied by C_n and added to the sinusoidal load.

Table 1. Dynamical beam characteristics for $n = 120;\ 320;\ 360;\ 400$.

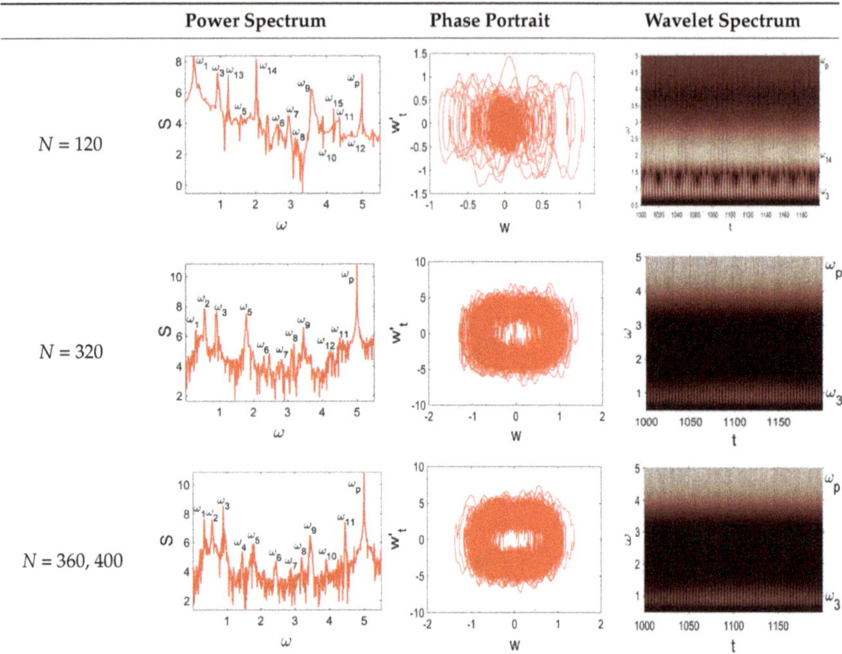

	Power Spectrum	Phase Portrait	Wavelet Spectrum
$N = 120$			
$N = 320$			
$N = 360, 400$			

Table 2. LLEs (the largest Lyapunov exponents) computed by the Wolf, Rosenstein and Kantz methods.

λ	q	n	Wolf	Rosenstein	Kantz
50	10000	160	0.10078	0.11334	0.04127
50	10000	200	0.07357	0.07359	0.02888
50	10000	240	0.09443	0.08599	0.03110
50	10000	360; 400	0.09885	0.07995	0.03190

Table 3 includes Morlet wavelet transform, Fourier spectrum, vibrational signal, Poincaré map, and 2D phase portraits for the beam partitions $n = \{80;\ 400\}$ (for fixed $C_n = 5000$). Results presented in Table 4 show that the most "noisy parts" of the Fourier and wavelet spectra were yielded by pink noise with dominating low-frequency band. The wavelet spectrum and the Fourier spectrum exhibit chaos on for all frequencies, phase portraits present dense clouds of phase plots, whereas Poincaré maps exhibit "chaotically" distributed points. Spectral densities of blue and white noise are similar, but white noise exhibits more low-frequency power, and hence the system is less chaotic in this case. On the contrary to the Fourier spectra, the phase portraits for white and brown noise are very similar. The strength of chaos again depends on low-frequency components of noise. This is validated by purple noise with dominating high-frequency components.

The methodology for finding the largest Lyapunov exponents by the Kantz and Rosenstein methods required the estimation of a tangent of the inclination angle of the most linear part of the time-dependent functions (Table 5). For this purpose, the least-squares algorithm was employed.

All possible points of the graph are taken for approximation at a given minimum length in time in percentage (in this case 30%) and the interval corresponding to the smallest sum of squares of the deviation of the initial data from this interval is selected.

The function $S(\tau)$ represents the dependence of the average divergence of the corresponding trajectories for the initial states governed by the inequality $\left|x(t) - x(t_j)\right| < \varepsilon$ (ε is a certain small quantity) for the considered time evolution τ.

Table 3. Characteristics of the beam vibrations ($Cn = 5000$, $\omega_0 = 1.0853$, $q_0 = 5000$).

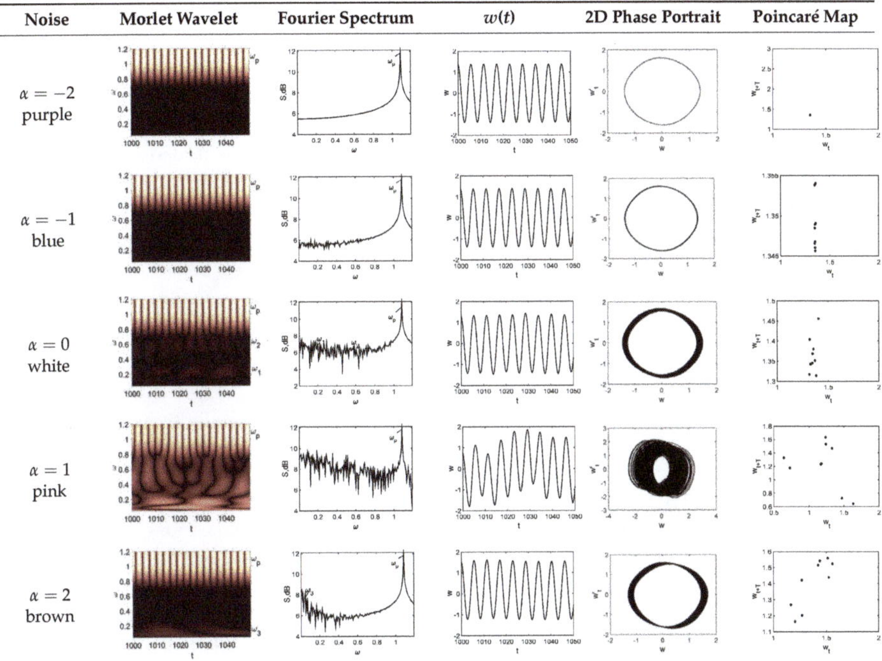

Table 4. Dependence of the average divergence of the neighborhood trajectories versus the considered evolution time τ.

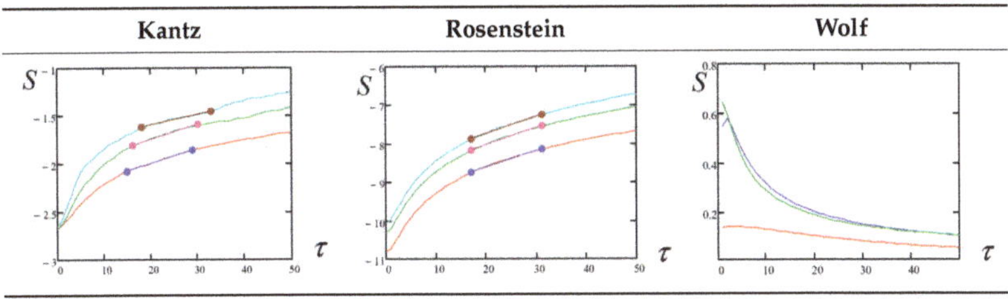

Table 5. LLEs for the investigated solutions (W/K/R state for the Wolf/Kantz/Rosenstein method, respectively).

$\alpha = -2$	$\alpha = -1$	$\alpha = 0$	$\alpha = 1$	$\alpha = 2$
W: 0.0	W: 0.0	W: 0.0	W: 0.0	W: 0.0
K: 0.0	K: $2 \cdot 10^{-2}$	K: $3 \cdot 10^{-2}$	K: $3 \cdot 10^{-2}$	K: $2 \cdot 10^{-2}$
R: 0.0	R: $6 \cdot 10^{-2}$	R: $6 \cdot 10^{-2}$	R: $7 \cdot 10^{-2}$	R: $6 \cdot 10^{-2}$

4. Concluding Remarks

The second part of this paper dealt with more complicated nonlinear dynamics exhibited by the Bernoulli–Euler beam subjected to deterministic and noisy input. In this case, the following main conclusions were yielded.

1. Convergence of the employed numerical methods was investigated with respect to the spatial and time coordinates (finite difference method with approximation $O(h^2)$ and the Runge–Kutta type methods). Numerical results showed that, to achieve reliable conclusions, it is necessary to conduct a complex analysis and, owing to the proposed methodology, each signal should be studied separately.

2. The most negligible effect was observed when purple noise was added—beam vibrations remained periodic. On the other hand, beam vibrations were significantly influenced by pink noise. The degree of chaotization of the system essentially depends on the presence of low-frequency components in noise. The employment of the Morlet mother wavelets allowed to detect time evolution of frequencies during chaotic beam vibrations (Table 2 ($\alpha = 0; 1; 2$)).

3. It was found, illustrated and discussed that the Wolf (W), Kantz (K) and Rosenstein (R) methods may yield significantly different values of Lyapunov exponents for the same signal. On the other hand, all the above-mentioned methods exhibited good correlation when used to study different colored noises.

4. The obtained results indicate a need to employ a more complex study by using qualitatively different numerical approaches to obtain reliable/true chaotic vibrations.

It can be concluded that, to obtain the most reliable results, it is necessary to consider a mechanical system (flexible beam) as a system with an infinite number of freedom degrees. One should investigate the convergence of the solution when reducing the partial differential equations to the Cauchy problem by the second-order finite element method $O(h^2)$. The number of the beam length should be maximized (in this work, from n = 50 to $n = 400$ partitions were considered). The Cauchy problem can be solved by several other methods, such as Runge–Kutta method. For every problem, the largest Lyapunov exponents need to be investigated in several ways (for instance, by Wolf, Kantz, and Rosenstein methods). Finally, the presented complex and diverse research study guaranteed the reliability of the obtained results.

Acknowledgments: This work has been supported by the grants the Russian Science Foundation, RSF 16-19-10290.

Author Contributions: Nikolay P. Erofeev, Vitalyj Dobriyan. and Marina A. Barulina performed the numerical study; Anton V. Krysko and Vadim A. Krysko analyzed the obtained results; and Jan Awrejcewicz wrote the paper.

Conflicts of Interest: The authors declare no conflict of interest.

References

1. Ruelle, D.; Takens, F. On the nature of turbulence. *Commun. Math Phys.* **1971**, *20*, 167–192. [CrossRef]
2. Berge, P.; Dubois, M.; Manneville, P.; Pomeau, Y. Intermittency in Rayleigh-Benard convection. *J. Phys. Paris Lett.* **1980**, *41*, L341–L345. [CrossRef]
3. Horsthemke, W.; Lefever, R. *Noise-Induced Transitions. Theory and Applications in Physics, Chemistry, and Biology*; Springer: Berlin, Germany, 1984.
4. Awrejcewicz, J.; Krysko, V.A.; Vakakis, A.F. *Nonlinear Dynamics of Continuous Elastic System*; Springer: Berlin, Germany, 2004.
5. Awrejcewicz, J.; Krysko, A.V.; Bochkarev, V.V.; Babenkova, T.V.; Papkova, I.V.; Mrozowski, J. Chaotic vibrations of two-layered beams and plates with geometric, physical and design nonlinearities. *Int. J. Bifurc. Chaos* **2011**, *21*, 2837–2851. [CrossRef]
6. Awrejcewicz, J.; Krysko, A.V.; Soldatov, V.; Krysko, V.A. Analysis of the nonlinear dynamics of the Timoshenko flexible beams using wavelets. *J. Comput. Nonlinear Dyn.* **2012**, *7*, 011005. [CrossRef]

7. Awrejcewicz, J.; Krysko, V.A.; Papkova, I.V.; Krysko, A.V. Routes to chaos in continuous mechanical systems. Part 1: Mathematical models and solution methods. *Chaos Solitons Fractals* **2012**, *45*, 687–708. [CrossRef]
8. Awrejcewicz, J.; Krysko, V.A.; Papkova, I.V.; Krysko, A.V. Routes to chaos in continuous mechanical systems. Part 2: Modeling transitions from regular to chaotic dynamics. *Chaos Solitons Fractals* **2012**, *45*, 709–720. [CrossRef]
9. Awrejcewicz, J.; Krysko, V.A.; Papkova, I.V.; Krysko, A.V. Routes to chaos in continuous mechanical systems. Part 3: The Lyapunov exponents, hyper, hyper-hyper and spatial-temporal chaos. *Chaos Solitons Fractals* **2012**, *45*, 721–736. [CrossRef]
10. Amabili, M. Nonlinear vibrations of doubly curved shallow shells. *Int. J. Non Linear Mech.* **2005**, *40*, 683–710. [CrossRef]
11. Amabili, M.; Sarkar, A.; Païdoussis, M.P. Chaotic vibrations of circular cylindrical shells: Galerkin versus reduced-order models via the proper orthogonal decomposition method. *J. Sound Vib.* **2006**, *290*, 736–762. [CrossRef]
12. Gulick, D. *Encounters with Chaos*; McGraw-Hill: New York, NY, USA, 1992.
13. Holmes, P.J. A nonlinear oscillator with a strange attractor. *Philos. Trans. R. Soc. A* **1979**, *292*, 419–448. [CrossRef]
14. Yee, H.; Sweby, C. Dynamical approach study of spurious steady state numerical solutions of nonlinear differential equations. II. Global asymptotic behavior of time discretizations. *Int. J. Comput. Fluid Dyn.* **1995**, *4*, 219–283. [CrossRef]
15. Corless, R.M.; Essex, C.; Nerenberg, M.A.H. Numerical methods can suppress chaos. *Phys. Lett. A* **1991**, *157*, 27–36. [CrossRef]
16. Tseng, W.Y.; Dugundji, J. Nonlinear vibrations of a buckled beam under harmonic excitation. *ASME J. Appl. Mech.* **1971**, *38*, 467–476. [CrossRef]
17. Moon, F.C.; Holmes, P.J. A magneto-elastic strange attractor. *J. Sound Vib.* **1979**, *65*, 275–296. [CrossRef]
18. Holmes, P.J.; Marsden, J. A partial differential equation with infinitely many periodic orbits: Chaotic oscillations of a forced beam. *Arch. Ration. Mech. Anal.* **1981**, *76*, 135–166. [CrossRef]
19. Szemplińska-Stupnicka, W.; Plaut, R.H.; Hsieh, J.C. Period doubling and chaos in unsymmetric structures under parametric excitation. *ASME J. Appl. Mech.* **1989**, *56*, 947–952. [CrossRef]
20. Bar-Yoseph, P.Z.; Fisher, D.; Gottlieb, O. Spectral element methods for nonlinear spatio-temporal dynamics of an Euler-Bernoulli beam. *Comput. Mech.* **1996**, *19*, 136–151. [CrossRef]
21. Dunne, J.F.; Ghanbari, M. Efficient extreme value prediction for nonlinear beam vibrations using measured random response histories. *Nonlinear Dyn.* **2001**, *24*, 71–101. [CrossRef]
22. Ge, G.; Xu, J. Stochastic bifurcation of one flexible beam subject to axial Gauss white noise excitation. In Proceedings of the 4th IEEE Conference on Industrial Electronics and Applications, Xi'an, China, 25–27 May 2009; pp. 1647–1653.
23. Cacciola, P.; Impollonia, N.; Muscolino, G. Crack detection and location in a damaged beam vibrating under white noise. *Comput. Struct.* **2003**, *81*, 1773–1782. [CrossRef]
24. Sahoo, B.; Panda, L.N.; Pohit, G. Nonlinear dynamics of an Euler-Bernoulli beam with parametric and internal resonances. *Procedia Eng.* **2013**, *64*, 727–736. [CrossRef]
25. Lan, C.B.; Qin, W.Y. Energy harvesting from coherent resonance of horizontal vibration of beam excited by vertical base motion. *Appl. Phys. Lett.* **2014**, *105*, 11391. [CrossRef]
26. Xu, L.; Ma, Q. Existence of random attractors for the floating beam equation with strong damping and white noise. *Bond. Value Probl.* **2015**, *2015*, 126. [CrossRef]
27. Devaney, R.L. *An Introduction to Chaotic Dynamical Systems*, 2nd ed.; Addison-Wesley: Reading, MA, USA, 1989.
28. Banks, J.; Brooks, J.; Davis, G.; Stacey, P. On Devaney's definition of chaos. *Am. Math. Mon.* **1992**, *99*, 332–334. [CrossRef]
29. Knudsen, C. Chaos without periodicity. *Am. Math. Mon.* **1994**, *101*, 563–565. [CrossRef]
30. Awrejcewicz, J.; Krysko, V.A.; Papkova, I.V.; Krysko, A.V. *Deterministic Chaos in One-Dimentional Continuous Systems*; World Scientific: Singapore, 2016.
31. Süli, E.; Mayers, D. *An Introduction to Numerical Analysis*; Cambridge University Press: Cambridge, UK, 2003.
32. Fehlberg, E. *Low-Order Classical Runge-Kutta Formulas with Step Size Control and Their Application to Some Heat Transfer Problems*; NASA Technical Report; NASA: Washington, WA, USA, 1969; p. 315.

33. Fehlberg, E. Klassische Runge-Kutta-Formeln vierter und niedrigerer Ordnung mit Schrittweiten-Kontrolle und ihre Anwendung auf Wärmeleitungsprobleme. *Computing* **1970**, *6*, 61–71. [CrossRef]

34. Cash, J.R.; Karp, A.N. A variable order Runge-Kutta method for initial value problems with rapidly varying right-hand sides. *ACM Trans. Math. Softw.* **1990**, *16*, 201–222. [CrossRef]

35. Dormand, J.R.; Prince, P.J. A family of embedded Runge-Kutta formulae. *J. Comput. Appl. Math.* **1980**, *6*, 19–26. [CrossRef]

36. Haar, A. Zur Theorie der orthogonalen Funktionensysteme. *Math. Ann.* **1910**, *69*, 331–371. [CrossRef]

37. Meyer, Y. *Ondelettes, Functions Splines et Analyses Graduees*; Lectures Given at the University of Torino: Torino, Italy, 1986.

38. Daubechies, I.; Grossmann, A.; Meyer, Y. Painless nonorthogonal expansions. *J. Math. Phys.* **1986**, *27*, 1271–1283. [CrossRef]

39. Grossman, A.; Morlet, S. Decomposition of Hardy functions into square separable wavelets of constant shape. *SIAM J. Math. Anal.* **1984**, *15*, 723–736. [CrossRef]

40. Meyer, Y. *Ondelettes et Fonctions Splines*; Technical Report; Seminaire EDP; Ecole Polytechnigue: Paris, France, 1986.

41. Meyer, Y. *Wavelets: Algorithms and Applications*; SIAM: Philadelphia, PA, USA, 1993.

42. Kantz, H. A robust method to estimate the maximum Lyapunov exponent of a time series. *Phys. Lett. A* **1994**, *185*, 77–87. [CrossRef]

43. Wolf, A.; Swift, J.B.; Swinney, H.L.; Vastano, J.A. Determining Lyapunov Exponents from a time series. *Phys. D Nonlinear Phenom.* **1985**, *16*, 285–317. [CrossRef]

44. Rosenstein, M.T.; Collins, J.J.; De Luca, C.J. A Practical Method for Calculating Largest Lyapunov Exponents from Small Data Sets. *Phys. D Nonlinear Phenom.* **1993**, *65*, 117–134. [CrossRef]

45. Volmir, A.S. *Nonlinear Dynamics of Plates and Shells*; Nauka: Moscow, Russian, 1972.

Article

On Points Focusing Entropy

Ewa Korczak-Kubiak *,†, Anna Loranty † and Ryszard J. Pawlak †

Faculty of Mathematics and Computer Science, Łódź University, Banacha 22, 90-238 Łódź, Poland;
loranta@math.uni.lodz.pl (A.L.); rpawlak@math.uni.lodz.pl (R.J.P.)
*   Correspondence: ekor@math.uni.lodz.pl
†   These authors contributed equally to this work.

Received: 22 January 2018; Accepted: 13 February 2018; Published: 16 February 2018

Abstract: In the paper, we consider local aspects of the entropy of nonautonomous dynamical systems. For this purpose, we introduce the notion of a (asymptotical) focal entropy point. The notion of entropy appeared as a result of practical needs concerning thermodynamics and the problem of information flow, and it is connected with the complexity of a system. The definition adopted in the paper specifies the notions that express the complexity of a system around certain points (the complexity of the system is the same as its complexity around these points), and moreover, the complexity of a system around such points does not depend on the behavior of the system in other parts of its domain. Any periodic system "acting" in the closed unit interval has an asymptotical focal entropy point, which justifies wide interest in these issues. In the paper, we examine the problems of the distortions of a system and the approximation of an autonomous system by a nonautonomous one, in the context of having a (asymptotical) focal entropy point. It is shown that even a slight modification of a system may lead to the arising of the respective focal entropy points.

Keywords: nonautonomous (autonomous) dynamical system; topological entropy; (asymptotical) focal entropy point; disturbation; m-dimensional manifold

MSC: 54C70; 37A35; 37B40; 58C30; 26A18

---

1. Introduction and Preliminaries

In many papers dealing with dynamical systems, their strong relation to difference equations is pointed out (see [1]), which gives the possibilities of their wide applications in many fields of knowledge, including economics, biology, information flow or physics [2–7]. Among the problems connected with "dynamical systems with discrete time observations", a special role is played by the entropy of these systems, which may be treated as a "measure" of the complexity of a dynamical system. This notion was introduced with respect to the issues connected with thermodynamics and the problem of "information loss" (more details on this topic can be found in [8]). At the beginning, the notion of entropy was related to the measure theory. Later, there appeared the notion of topological entropy introduced by R. Adler, A. Konheim and J. McAndrew [9], and next, an equivalent definition for metric spaces was formulated [10,11]. It is worth mentioning that in the further stage of research, the definition of topological entropy for discontinuous functions was also studied [12]. The considerations mentioned concerned autonomous systems. Later, still, there appeared results regarding the entropy of nonautonomous dynamical systems. We will base our investigations, among others, on [13]. In general, the notion of entropy concerns a global property of dynamical systems. However, research connected for example with stability points or non-wandering points, as well as the analysis of various examples of functions lead to the conclusion that it is also purposeful to examine local aspects of entropy and points around which the entropy is "focused" in some sense, e.g., [14,15]. Simultaneously, the example presented in [16] (p. 1118) shows that it is intentional to assume that the essence of a point "focusing" entropy should be connected with the behavior of functions only (exclusively) around this point or the value of functions at this point (sometimes, the fact that a point is a full entropy point [15] is decisively influenced by the behavior of a function "far away" from that point). For that reason, a new approach to this problem was introduced in [16]. All the above-mentioned papers concerned autonomous systems. In this paper, we will refer to these issues as nonautonomous dynamical systems. Our considerations will be mainly connected with the periodicity of the examined systems. Such kind of investigations are frequently connected in the literature with such systems (e.g. [6,13,17]). It is caused by the connections of such systems with periodic difference equations (it is well signalled in [3]).

Throughout the paper, the symbol $\mathbb{N}$ will stand for the set of all positive integers and $\mathbb{N}_0 = \mathbb{N} \cup \{0\}$. Moreover, (X, ρ) will denote a compact metric space. The closure, the interior and the cardinality of a set $A \subset X$ will be denoted by $\mathrm{cl}(A)$, $\mathrm{int}(A)$ and $\#(A)$, respectively. For any function $f : X \to X$ and sets $A, B \subset X$, the symbols $f \upharpoonright A$ and $A \underset{f}{\to} B$ mean the restriction of f to A and $B \subset f(A)$, respectively.

The symbol $\mathrm{FIX}_X(x_0)$ will denote the family of all self-maps defined on X such that the point x_0 is their fixed point, and the symbol $\mathrm{FIX}(f)$ will stand for the set of all fixed points of a function f. Moreover, for any functions $f, g : X \to X$, let us adopt the following notation $\neq (f, g) = \{x \in X : f(x) \neq g(x)\}$.

Let (X, ϱ) be a metric space and $\{K_n\}_{n \in \mathbb{N}}$ be a sequence of nonempty closed subsets of X. We shall say that the sequence $\{K_n\}_{n \in \mathbb{N}}$ has the extension property if for any $i, j \in \mathbb{N}$ and any continuous function $\varphi : A \to K_j$, where $A \subset K_i$ is a closed set, one can find a continuous function $\psi : K_i \to K_j$, which is an extension of φ, i.e., $\psi \upharpoonright A = \varphi$. Obviously, if for example $X = R^n$ and K_n are cubes, then this fact follows from the generalizations of the classical Tietze theorem.

Following [13], by a nonautonomous dynamical system on X (NDS), we will mean any sequence of functions $f_{1,\infty} = \{f_i\}_{i \in \mathbb{N}}$ such that $f_i : X \to X$. If $f_i = f$ for $i \in \mathbb{N}$, then we call the system autonomous and denote it by (f). For $n \in \mathbb{N}$, let $f_{n,\infty} = \{f_n, f_{n+1}, \dots\}$ and $f_{1,\infty}^n = \{f_{(i-1) \cdot n+1}^n\}_{i \in \mathbb{N}}$, where $f_i^n = f_{n+i-1} \circ f_{n+i-2} \circ \cdots \circ f_{i+1} \circ f_i$. Moreover, let $f_i^0 = f_i^{-0} = \mathrm{id}$ (where id is the identity function) and $f_i^{-n} = f_i^{-1} \circ f_{i+1}^{-1} \circ \cdots \circ f_{i+(n-1)}^{-1}$ for any $i, n \in \mathbb{N}$ (the last notation will be applied to sets, so we do not assume that these maps are invertible). If $f : X \to X$ is a function, then for any $n \in \mathbb{N}$, the symbol f^n will denote the n-th iteration of f, i.e., $f^n = f \circ f^{n-1}$ and $f^0 = \mathrm{id}$.

We say that a dynamical system $f_{1,\infty}$ is periodic with a period n if $f_k = f_{k \bmod n}$, if $k \bmod n \neq 0$ and $f_k = f_n$ otherwise. Moreover, we say that x_0 is a periodic point with a period n of an NDS $f_{1,\infty}$ if x_0 is a fixed point of an NDS $f_{1,\infty}^n$, i.e., $f_{(i-1) \cdot n+1}^n(x_0) = x_0$ for any $i \in \mathbb{N}$.

If M is a matrix, then the trace of M will be denoted by $\mathrm{tr}(M)$. Let $\{M_n\}_{n \in \mathbb{N}}$ be a sequence of square matrices of the same degree t. Then, for any $k \in \mathbb{N}$, we will consider $\prod_{i=1}^{k} M_i = M_1 \cdot M_2 \cdot \cdots \cdot M_k$.

In [13] was introduced a Bowen-like definition of entropy for an NDS consisting of continuous functions. This definition was expanded for systems consisting of arbitrary functions in the paper [8]. We will briefly review that notion.

Let $n \in \mathbb{N}$ and $\varepsilon > 0$. A set $E \subset X$ is called (n, ε)-separated if for any two distinct points $x, y \in E$, there exists $j \in \{0, \dots, n-1\}$ such that $\rho(f_1^j(x), f_1^j(y)) > \varepsilon$. If $Y \subset X$, then E is (n, ε)-separated in Y if it satisfies the above condition and $E \subset Y$. Let $s_n(f_{1,\infty}, Y, \varepsilon)$ denote the maximal cardinality of the (n, ε)-separated set in Y. Then, the entropy of a system $f_{1,\infty}$ on Y is the number:

$$h(f_{1,\infty}, Y) = \lim_{\varepsilon \to 0} \limsup_{n \to \infty} \frac{1}{n} \log s_n(f_{1,\infty}, Y, \varepsilon).$$

If $Y = X$, then we write briefly $h(f_{1,\infty})$ instead of $h(f_{1,\infty}, X)$. Moreover, if we consider an autonomous system (f), then the entropy of this system will be denoted by $h(f, Y)$ and $h(f)$, respectively. By the entropy of a function f, we will mean the entropy of a respective autonomous system (f).

Now, we will signal, in the form of lemmas, basic facts that will be used in the further part of the paper. Reasoning similar to that in the proofs of Lemma 4.3 and 4.5 [13] allows proving the following result concerning the entropy of an NDS consisting of not necessarily continuous functions.

Lemma 1. *Let $f_{1,\infty}$ be a dynamical system. Then, for any $n \geq 1$, we have:*

$$h(f_{1,\infty}^n) \leq n \cdot h(f_{1,\infty}).$$

Lemma 2. *Let $f_{1,\infty}$ be a dynamical system on X. For any $1 \leq i \leq j < \infty$, we have $h(f_{i,\infty}) \leq h(f_{j,\infty})$.*

In the case of NDS, entropy does not always fully reflect the complexity of a system (see, e.g., the considerations in [13], p. 216). Therefore, in [13] was introduced a new notion of asymptotical entropy, which, with respect to autonomous systems, coincides with the classical entropy.

An asymptotical entropy of a dynamical system $f_{1,\infty}$ is the number $h^*(f_{1,\infty})$ defined as follows: $h^*(f_{1,\infty}) = \lim_{n\to\infty} h(f_{n,\infty})$. The existence of such a limit follows from Lemma 2. Moreover, Lemma 2 allows concluding that $h(f_{1,\infty}) \leq h^*(f_{1,\infty})$. It is worth adding that the inequality from Lemma 2 is not true for entropy on subsets of the space, so the asymptotical entropy of a system on a set $Y \subset X$ is defined as the following upper limit:

$$h^*(f_{1,\infty}, Y) = \limsup_{n\to\infty} h(f_{n,\infty}, Y).$$

Our terminology and notations related to m-dimensional manifolds will coincide with those of [18]. An m-dimensional manifold with a boundary is a nonempty compact metric space $(\mathbb{M}, d)$ such that every point $q \in \mathbb{M}$ has a neighborhood U that is homeomorphic (via a transformation called the chart on U) to an open subset of the m-dimensional upper half space $\mathbb{H}^m = \{(x_1, ..., x_m) \in \mathbb{R}^m : x_m \geq 0\}$. Since any open ball in $\mathbb{R}^m$ is homeomorphic to some open subset of $\mathbb{H}^m$, an m-dimensional topological manifold is an m-dimensional topological manifold with a boundary (with an empty boundary). Therefore, in this paper, we will consider only m-dimensional topological manifolds with a boundary.

If $\mathbb{M}$ is a nonempty m-dimensional manifold with a boundary, a point that belongs to the inverse image of $\mathrm{int}(\mathbb{H}^m) = \{(x_1, ..., x_m) \in \mathbb{R}^m : x_m > 0\}$ under some chart is called an interior point of $\mathbb{M}$. The set of all interior points of a manifold $\mathbb{M}$ will be denoted by $\mathrm{Int}(\mathbb{M})$. The symbol $\mathfrak{B}_\mathbb{M}$ will stand for the set of all closed submanifolds $\mathcal{M}$ of $\mathbb{M}$ (i.e., $\mathcal{M} \subset \mathbb{M}$ is a closed manifold) such that the dimensions of $\mathcal{M}$ and $\mathbb{M}$ are the same.

We shall say that an NDS $(f_{1,\infty})$ of functions defined on $\mathbb{M}$ is irreducible at x_0 if for $n \in \mathbb{N}$, a function f_1^n is irreducible at x_0, i.e., for any open neighborhood U of x_0, there exists a point $y_0 \in \mathrm{Int}(\mathbb{M}) \cap U$ such that $f_1^n(x_0) \neq f_1^n(y_0)$.

2. Focal Entropy Points of NDS

Now, we will introduce the notion of a focal entropy point of NDS, having in mind the general assumption: the fact that a given point is a focal entropy point means that the complexity of the system in any neighborhood of this point is the same as the complexity of the whole system and does not depend on the behavior of functions around other points.

Let $\mathfrak{A}$ be a family of nonempty subsets of X such that each nonempty open set contains some element of $\mathfrak{A}$. In view of the considerations presented in this paper, from now on, we will assume that $\mathfrak{A}$ contains the family of all closed sets of cardinality continuum.

Put $\Theta(\mathfrak{A}) = \{(A_1, ..., A_m) : A_1, ..., A_m \in \mathfrak{A}, m \in \mathbb{N}, \mathrm{cl}(A_i) \cap \mathrm{cl}(A_j) = \emptyset \text{ for } i \neq j\}$.

Let $\mathcal{A} = (A_1, ..., A_m) \in \Theta(\mathfrak{A})$ and $n \in \mathbb{N}$. Set $M_{f_n}(\mathcal{A}) = [a_{i,j}^{f_n}]_{i,j \leq m}$, where:

$$a_{i,j}^{f_n} = \begin{cases} 1 & \text{if } A_i \underset{f_n}{\to} A_j, \\ 0 & \text{if } A_j \setminus f_n(A_i) \neq \emptyset. \end{cases}$$

Moreover, for $k \in \mathbb{N}$, a system $(f_{1,\infty})$ and $\mathcal{A} = (A_1, \ldots, A_m) \in \Theta(\mathfrak{A})$, let:

$$M^k_{f_{1,\infty}}(\mathcal{A}) = \prod_{n=1}^{k} M_{f_n}(\mathcal{A}) := [a^k_{i,j}]_{i,j \leq m}. \tag{1}$$

The entropy of $f_{1,\infty}$ with respect to the sequence $\mathcal{A}$ is the following number:

$$H_{f_{1,\infty}}(\mathcal{A}) = \begin{cases} \limsup_{k \to \infty} \frac{1}{k} \log \operatorname{tr}(M^k_{f_{1,\infty}}(\mathcal{A})) & \text{if } \operatorname{tr}(M^k_{f_{1,\infty}}(\mathcal{A})) > 0, \\ 0 & \text{if } \operatorname{tr}(M^k_{f_{1,\infty}}(\mathcal{A})) = 0. \end{cases}$$

The process of computing the entropy of a system with respect to a sequence of sets may be simplified by introducing the notion of a path. Let $k \in \mathbb{N}$. For a k-path connected with the sequence $\mathcal{A}$ and with the dynamical system $(f_{1,\infty})$, we call each sequence of sets $(A_{p_1}, A_{p_2}, \ldots, A_{p_k})$ such that $p_i \in \{1, \ldots, m\}$ for $i = 1, \ldots, k$ and:

$$A_{p_1} \xrightarrow{f_1} A_{p_2} \xrightarrow{f_2} A_{p_3} \xrightarrow{f_3} \cdots \xrightarrow{f_{k-2}} A_{p_{k-1}} \xrightarrow{f_{k-1}} A_{p_k}.$$

The sets $A_{p_1}, A_{p_2}, \ldots, A_{p_k}$ are called the nodes of the path. If no confusion can arise, we will write simply k-path. We say that a point $x_0 \in A_{p_1}$ is connected with a k-path $(A_{p_1}, A_{p_2}, \ldots, A_{p_k})$ if $f_1^i(x_0) \in A_{p_{i+1}}$ for $i = 1, \ldots, k-1$. It is easy to see that such a point exists for any path.

One can easily notice that the entry $a^k_{i,j}$, where $i, j \in \{1, \ldots, m\}$, of the matrix (1) is equal to the number of $(k+1)$-paths connected with the sequence $\mathcal{A}$ and the NDS $f_{1,\infty}$ such that the set A_i is the first node of the path and the set A_j is its last node. Consequently, $\operatorname{tr}(M^k_{f_{1,\infty}}(\mathcal{A}))$ is equal to the number of $(k+1)$-paths connected with the sequence $\mathcal{A}$ and the NDS $f_{1,\infty}$ such that the set A_i is simultaneously the first and the last node of the path, for $i = 1, \ldots, m$.

Now, let us state the theorem, which will allow introducing the next steps of the definition.

Theorem 1. *Let $f_{1,\infty}$ be an NDS, $\mathcal{A} = (A_1, \ldots, A_m) \in \Theta(\mathfrak{A})$ and $n \in \mathbb{N}$. Then:*

$$H_{f_{1,\infty}^n}(\mathcal{A}) \leq h(f_{1,\infty}^n) \leq n \cdot h(f_{1,\infty}).$$

Proof. The second inequality follows from Lemma 1, so it is sufficient to show the first inequality. Suppose, contrary to our claim, that there exists a real number α such that:

$$h(f_{1,\infty}^n) < \alpha < H_{f_{1,\infty}^n}(\mathcal{A}). \tag{2}$$

It is obvious that $\alpha > 0$ and $H_{f_{1,\infty}^n}(\mathcal{A}) > 0$. According to our assumptions connected with the family $\Theta(\mathfrak{A})$, we have $\varepsilon_{\mathcal{A}} = \frac{1}{2} \min\{\rho(\operatorname{cl}(A_i), \operatorname{cl}(A_j)) : i, j \in \{1, \ldots, m\} \wedge i \neq j\} > 0$. Taking into account (2), we obtain that there exists an increasing sequence of positive integers $\{k_s\}_{s \in \mathbb{N}}$ such that:

$$\frac{1}{k_s} \log \operatorname{tr}(M^{k_s}_{f_{1,\infty}^n}(\mathcal{A})) > \alpha \text{ for } s = 1, 2, \ldots \tag{3}$$

Clearly, $a^{k_s}_{1,1}, a^{k_s}_{2,2}, \ldots, a^{k_s}_{m,m}$ are successive entries of the main diagonal of the matrix $M^{k_s}_{f_{1,\infty}^n}(\mathcal{A})$, for any $s \in \mathbb{N}$. By (3), one can conclude that for any $s \in \mathbb{N}$, we have $N_{k_s} = \{i \in \{1, \ldots, m\} : a^{k_s}_{i,i} > 0\} \neq \varnothing$. For any $s \in \mathbb{N}$ and $i \in \{1, \ldots, m\}$, the number of $(k_s + 1)$-paths of the form $A_i \xrightarrow{f_1^n} A_{p_1} \xrightarrow{f_{n+1}^n}$

$A_{p_2} \xrightarrow{f_{2n+1}^n} \cdots \xrightarrow{f_{(k_s-2)n+1}^n} A_{p_{k_s-1}} \xrightarrow{f_{(k_s-1)n+1}^n} A_i$, where $p_w \in \{1, \ldots, m\}$ for $w = 1, \ldots, k_s - 1$, is equal to $a^{k_s}_{i,i}$.

For any $s \in \mathbb{N}$ and $i \in N_{k_s}$, let $\beta^{k_s}_i$ denote the set of all $(k_s + 1)$-paths whose first and last node is A_i. Obviously $\#(\beta^{k_s}_i) = a^{k_s}_{i,i}$. Therefore, let $\beta^{k_s}_i = \{B^{k_s}_{i,1}, B^{k_s}_{i,2}, \ldots, B^{k_s}_{i,a^{k_s}_{i,i}}\}$. For any $s \in \mathbb{N}$, $i \in N_{k_s}$ and $j \in \{1, \ldots, a^{k_s}_{i,i}\}$, let $b^{k_s}_{i,j}$ be a point connected with the path $B^{k_s}_{i,j}$.

Put $\Delta(k_s) = \{b_{i,j}^{k_s} : i \in N_{k_s} \wedge j \in \{1,\ldots,a_{i,i}^{k_s}\}\}$ for $s \in \mathbb{N}$. It is easily seen that $b_{i,j}^{k_s} \in A_i$ for $s \in \mathbb{N}$, $i \in N_{k_s}$ and $j \in \{1,\ldots,a_{i,i}^{k_s}\}$. Thus, if $i_1, i_2 \in N_{k_s}$ and $i_1 \neq i_2$, then $b_{i_1,j}^{k_s} \neq b_{i_2,j}^{k_s}$. Moreover, if $j_1, j_2 \in \{1,\ldots,a_{i,i}^{k_s}\}$ and $j_1 \neq j_2$, then $b_{i,j_1}^{k_s} \neq b_{i,j_2}^{k_s}$. Thus, $\#(\Delta(k_s)) = \sum_{i \in N_{k_s}} \#(\beta_i^{k_s}) = \sum_{i \in N_{k_s}} a_{i,i}^{k_s}$, and finally,

$$\#(\Delta(k_s)) = \sum_{i=1}^{m} a_{i,i}^{k_s}, \text{ because } a_{i,i}^{k_s} = 0 \text{ for } i \in \{1,\ldots,m\} \setminus N_{k_s}.$$

Let $b_{i_1,j_1}^{k_s}, b_{i_2,j_2}^{k_s}$ be any distinct points of the set $\Delta(k_s)$. If $i_1 \neq i_2$, then $\rho(b_{i_1,j_1}^{k_s}, b_{i_2,j_2}^{k_s}) \geq \rho(\mathrm{cl}(A_{i_1}),\mathrm{cl}(A_{i_2})) > \varepsilon_A$. If $i_1 = i_2 = i$, then $j_1 \neq j_2$. Thus, since $b_{i,j_1}^{k_s}$ is connected with the path $B_{i,j_1}^{k_s} = (A_{i,j_1}, A_{p_1,j_1},\ldots, A_{p_{k_s-1},j_1}, A_{i,j_1})$ and $b_{i,j_2}^{k_s}$ is connected with the path $B_{i,j_2}^{k_s} = (A_{i,j_2}, A_{p_1,j_2},\ldots, A_{p_{k_s-1},j_2}, A_{i,j_2})$ and $B_{i,j_1}^{k_s} \neq B_{i,j_2}^{k_s}$, so there exists $w_0 \in \{1,\ldots,k_s-1\}$ such that $A_{p_{w_0},j_1} \neq A_{p_{w_0},j_2}$ and $\rho(f^{w_0 \cdot n}(b_{i,j_1}^{k_s}), f^{w_0 \cdot n}(b_{i,j_2}^{k_s})) \geq \rho(\mathrm{cl}(A_{p_{w_0},j_1}),\mathrm{cl}(A_{p_{w_0},j_2})) > \varepsilon_A$. This gives that $\Delta(k_s)$ is the (k_s, ε_A)-separated set for the system $(f_{1,\infty}^n)$.

As a consequence, we obtain $s_{k_s}(f_{1,\infty}^n, \varepsilon_A) \geq \#(\Delta(k_s)) = a_{1,1}^{k_s} + \cdots + a_{m,m}^{k_s}$. Let $\varepsilon \in (0, \varepsilon_A)$. Thus, $\limsup_{l \to \infty} \frac{1}{l} \log s_l(f_{1,\infty}^n, \varepsilon) \geq \limsup_{s \to \infty} \frac{1}{k_s} \log \sum_{i=1}^{m} a_{i,i}^{k_s} = \limsup_{s \to \infty} \frac{1}{k_s} \log \mathrm{tr}(M_{f_{1,\infty}^n}^{k_s}(A)) \geq \alpha$, and hence, $h(f_{1,\infty}^n) = \lim_{\varepsilon \to 0} \limsup_{l \to \infty} \frac{1}{l} \log s_l(f_{1,\infty}^n, \varepsilon) \geq \alpha$, which contradicts (2). $\square$

We continue the considerations leading to the definition of a focal entropy point. Let $U \subset X$ be an open set. For $\mathcal{A} = (A_1,\ldots,A_m) \in \Theta(\mathfrak{A})$, the notation $\mathcal{A} \subset U$ will mean that $A_i \subset U$ for any $i \in \{1,\ldots,m\}$. Let us adopt the following notation:

$$H(\mathfrak{A}, f_{1,\infty}, U) = \sup\left\{\frac{1}{n} H_{f_{1,\infty}^n}(\mathcal{A}) : \mathcal{A} \in \Theta(\mathfrak{A}) \wedge \mathcal{A} \subset U \wedge n \in \mathbb{N}\right\}.$$

Notice that on account of Theorem 1, for any open set U, we have:

$$H(\mathfrak{A}, f_{1,\infty}, U) \leq h(f_{1,\infty}). \tag{4}$$

Put:

$$d(\mathfrak{A}, f_{1,\infty}, U) = \begin{cases} \frac{H(\mathfrak{A}, f_{1,\infty}, U)}{h(f_{1,\infty})} & \text{if } h(f_{1,\infty}) \in (0,\infty), \\ 1 & \text{if } H(\mathfrak{A}, f_{1,\infty}, U) = \infty \text{ or } h(f_{1,\infty}) = 0, \\ 0 & \text{if } H(\mathfrak{A}, f_{1,\infty}, U) \in [0,\infty) \text{ and } h(f_{1,\infty}) = \infty. \end{cases}$$

Using the last quantity, one can define the next one in the following way:

$$E(\mathfrak{A}, f_{1,\infty}, x_0) = \inf\{d(\mathfrak{A}, f_{1,\infty}, U) : U \in O(x_0)\},$$

where $O(x_0)$ denotes the family of all open sets containing x_0.

According to Theorem 1, we have $E(\mathfrak{A}, f_{1,\infty}, x_0) \leq 1$. If $E(\mathfrak{A}, f_{1,\infty}, x_0) = 1$, then we say that a point $x_0 \in X$ is a $\mathfrak{A}$-focal entropy point of a system $f_{1,\infty}$.

Notice that if a system $f_{1,\infty}$ is autonomous, i.e., $f_i = f$ for $i \in \mathbb{N}$, then the definition of a $\mathfrak{A}$-focal entropy point of the system $f_{1,\infty}$ coincides with the definition introduced in [16].

If in the definition of the quantity $d(\mathfrak{A}, f_{1,\infty}, U)$ we will replace an entropy $h(f_{1,\infty})$ with asymptotical entropy $h^*(f_{1,\infty})$, then by defining in an analogous way as above, we will obtain the notion of a asymptotical $\mathfrak{A}$-focal entropy point of $f_{1,\infty}$. In such a case, we will use a star in the respective symbols: $d^*(\mathfrak{A}, f_{1,\infty}, U)$, $E^*(\mathfrak{A}, f_{1,\infty}, x_0)$. Therefore, we say that a point $x_0 \in X$ is an asymptotical $\mathfrak{A}$-focal entropy point of a system $f_{1,\infty}$ if $E^*(\mathfrak{A}, f_{1,\infty}, x_0) = 1$.

It is easy to see that if x_0 is an asymptotical $\mathfrak{A}$-focal entropy point of a system $f_{1,\infty}$, then it is a $\mathfrak{A}$-focal entropy point of this system. Obviously, if $f_{1,\infty}$ is periodic, then the notions of an asymptotical $\mathfrak{A}$-entropy point of the system and of a $\mathfrak{A}$-focal entropy point of the system coincide.

The natural question arises whether there exist such points. The next theorem is a partial answer to this problem.

Theorem 2. *Let $f_{1,\infty}$ be a periodic dynamical system on $[0,1]$ consisting of continuous functions. Then, there exists an asymptotical $\mathfrak{A}$-focal entropy point of the system $f_{1,\infty}$.*

Proof. Let n be a period of the system $f_{1,\infty}$. Put $f = f_1^n$ and $g_{1,\infty} = f_{1,\infty}^n$. Then, $g_{1,\infty} = (f)$. Hence, by Lemma 1, we obtain:

$$h(g_{1,\infty}) = n \cdot h(f_{1,\infty}). \tag{5}$$

Moreover, notice that:

$$g_{1,\infty}^k = f_{1,\infty}^{n \cdot k} \text{ for } k \in \mathbb{N}. \tag{6}$$

By Corollary 4.5 [16], there exists a point $x_0 \in [0,1]$, which is a $\mathfrak{A}$-focal entropy point of $g_{1,\infty}$.

We will show that x_0 is a $\mathfrak{A}$-focal entropy point of the system $f_{1,\infty}$. Let $U \in O(x_0)$. It is easy to observe that $E(\mathfrak{A}, g_{1,\infty}, x_0) = 1$ and consequently $d(\mathfrak{A}, g_{1,\infty}, U) = 1$. We need to consider the following cases (we omit the trivial case $h(g_{1,\infty}) = 0$):

(i) $H(\mathfrak{A}, g_{1,\infty}, U) = \infty$. Thus, $\sup\{\frac{1}{k}H_{g_{1,\infty}^k}(\mathcal{A}) : \mathcal{A} \in \Theta(\mathfrak{A}) \wedge \mathcal{A} \subset U \wedge k \in \mathbb{N}\} = \infty$. For any $\beta > 0$, there exist $k_\beta \in \mathbb{N}$ and $\mathcal{A}_\beta \in \Theta(\mathfrak{A})$ such that $\mathcal{A}_\beta \subset U$ and $\frac{1}{k_\beta}H_{g_{1,\infty}^{k_\beta}}(\mathcal{A}_\beta) > n \cdot \beta$. Obviously, by (6), we have $g_{1,\infty}^{k_\beta} = f_{1,\infty}^{n \cdot k_\beta}$, so $\frac{1}{k_\beta}H_{f_{1,\infty}^{n \cdot k_\beta}}(\mathcal{A}_\beta) > n \cdot \beta$, and therefore, $\frac{1}{n \cdot k_\beta}H_{f_{1,\infty}^{n \cdot k_\beta}}(\mathcal{A}_\beta) > \beta$. As a consequence, $\sup\{\frac{1}{s}H_{f_{1,\infty}^s}(\mathcal{A}) : \mathcal{A} \in \Theta(\mathfrak{A}) \wedge \mathcal{A} \subset U \wedge s \in \mathbb{N}\} > \beta$. Hence and from arbitrariness β, we conclude that $\sup\{\frac{1}{s}H_{f_{1,\infty}^s}(\mathcal{A}) : \mathcal{A} \in \Theta(\mathfrak{A}) \wedge \mathcal{A} \subset U \wedge s \in \mathbb{N}\} = \infty$, and consequently, $d(\mathfrak{A}, f_{1,\infty}, U) = 1$.

(ii) $h(g_{1,\infty}) \in (0, \infty)$ and $H(\mathfrak{A}, g_{1,\infty}, U) = h(g_{1,\infty})$. By (5), we obtain $h(f_{1,\infty}) \in (0, \infty)$. We have $h(g_{1,\infty}) = \sup\{\frac{1}{k}H_{g_{1,\infty}^k}(\mathcal{A}) : \mathcal{A} \in \Theta(\mathfrak{A}) \wedge \mathcal{A} \subset U \wedge k \in \mathbb{N}\}$, so for any $\beta > 0$, there exist $k_\beta \in \mathbb{N}$ and $\mathcal{A}_\beta \in \Theta(\mathfrak{A})$ such that $\mathcal{A}_\beta \subset U$ and $\frac{1}{k_\beta}H_{g_{1,\infty}^{k_\beta}}(\mathcal{A}_\beta) > h(g_{1,\infty}) - n \cdot \beta$. Clearly, by (6), we may infer that $g_{1,\infty}^{k_\beta} = f_{1,\infty}^{n \cdot k_\beta}$, so $\frac{1}{k_\beta}H_{f_{1,\infty}^{n \cdot k_\beta}}(\mathcal{A}_\beta) > h(g_{1,\infty}) - n \cdot \beta$. By use of (5), we get $\frac{1}{n \cdot k_\beta}H_{f_{1,\infty}^{n \cdot k_\beta}}(\mathcal{A}_\beta) > h(f_{1,\infty}) - \beta$. Finally, we have shown that for any $\beta > 0$, there exist $l_\beta = n \cdot k_\beta \in \mathbb{N}$ and $\mathcal{A}_\beta \in \Theta(\mathfrak{A})$ such that $\mathcal{A}_\beta \subset U$ and $\frac{1}{l_\beta}H_{f_{1,\infty}^{l_\beta}}(\mathcal{A}_\beta) > h(f_{1,\infty}) - \beta$, so:

$$\sup\{\frac{1}{k}H_{f_{1,\infty}^k}(\mathcal{A}) : \mathcal{A} \in \Theta(\mathfrak{A}) \wedge \mathcal{A} \subset U \wedge k \in \mathbb{N}\} \geq h(f_{1,\infty}). \tag{7}$$

Moreover, according to (4), we have:

$$H(\mathfrak{A}, f_{1,\infty}, U) \leq h(f_{1,\infty}). \tag{8}$$

Finally, (7) and (8) give $H(\mathfrak{A}, f_{1,\infty}, U) = \sup\{\frac{1}{k}H_{f_{1,\infty}^k}(\mathcal{A}) : \mathcal{A} \in \Theta(\mathfrak{A}) \wedge \mathcal{A} \subset U \wedge k \in \mathbb{N}\} = h(f_{1,\infty})$. Thus, $d(\mathfrak{A}, f_{1,\infty}, U) = 1$.

Since U was chosen arbitrarily, we obtain $E(\mathfrak{A}, f_{1,\infty}, x_0) = 1$, so x_0 is a $\mathfrak{A}$-focal entropy point of $f_{1,\infty}$, and simultaneously, it is its asymptotical $\mathfrak{A}$-focal entropy point because this system is periodic. $\square$

3. Disturbance and Approximation

In various considerations connected with autonomous and nonautonomous dynamical systems, a special role is played by fixed points of the systems (e.g., stable points [6]). It is not difficult to find an example showing that a fixed point of NDS need not be its focal entropy point. On the other hand, a given NDS can be approximated or disturbed by entering new functions into it. In each of these operations, it is important to do it by means of functions that are close to the base NDS and belong to the common structure. This leads in a natural way to distinguishing equivalence classes.

Let $f, g \in \mathrm{FIX}_X(x_0)$ and $\varepsilon > 0$. In the set $\mathrm{FIX}_X(x_0)$, we will define the following relation:

$$f \frac{\varepsilon}{x_0} g \Leftrightarrow \neq (f, g), f(\neq (f, g)), g(\neq (f, g)) \subset B(x_0, \varepsilon), \tag{9}$$

where $B(x_0, \varepsilon)$ is an open ball with radius ε and center x_0. It is not difficult to show that for the fixed $\varepsilon > 0$ and $x_0 \in X$, the relation (9) is an equivalence relation in $\mathrm{FIX}_X(x_0)$.

The symbol $[f]_{x_0}^{\varepsilon}$ will stand for the equivalence class of $f \in \mathrm{FIX}_X(x_0)$ under the relation $\frac{\varepsilon}{x_0}$.

In this paper are mainly examined periodic dynamical systems, so it is natural to consider periodic disruptions called disturbances. The idea of the disturbance is introducing, in equal periods of time, a function belonging to the equivalence class generated by the iteration of functions lying between successive disturbance periods.

Let $f_{1,\infty}$ be a periodic NDS with a period $k_0 \in \mathbb{N}$, and let $\varepsilon > 0$. We say that $T_{1,\infty}^{\varepsilon}$ is a periodic ε-disturbance of $f_{1,\infty}$ if there exists a continuous function ψ such that:

(PD1) $T_{1,\infty}^{\varepsilon} = \{f_1, f_2, \dots, f_{k_0}, \psi, f_1, f_2, \dots, f_{k_0}, \psi, \dots\}$,
(PD2) $\psi \in [f_1^{k_0}]_{x_0}^{\varepsilon}$.

The next theorem shows that a periodic dynamical system may be periodically disturbed by means of a function belonging to an earlier defined equivalence class (with arbitrary small ε) in such a way that a periodic point of the system becomes its asymptotical $\mathfrak{A}$-focal entropy point.

Theorem 3. *Let $f_{1,\infty}$ be a periodic dynamical system on $\mathbb{M}$ consisting of continuous functions such that $x_0 \in \mathbb{M}$ is a periodic point of this NDS and $f_{1,\infty}$ is irreducible at x_0. For any $\varepsilon > 0$, there exists a system $T_{1,\infty}^{\varepsilon}$ that is a periodic ε-disturbance of $f_{1,\infty}$ such that x_0 is an asymptotical $\mathfrak{A}$-focal entropy point of $T_{1,\infty}^{\varepsilon}$.*

Proof. Let m_0 be a period of $f_{1,\infty}$ and m_1 be a period of x_0. Put $n_0 = m_0 \cdot m_1$. It follows immediately that n_0 is both a period of $f_{1,\infty}$ and of x_0. Let $\varepsilon > 0$ and $\{\mathcal{M}_n\}_{n=0}^{\infty} \subset \mathfrak{B}_{\mathbb{M}}$ be a sequence of connected submanifolds satisfying the following properties:

[M1] $x_0 \in \mathcal{M}_{n+1} \subset \mathrm{int}(\mathcal{M}_n)$ for $n \in \mathbb{N}_0$,
[M2] $f_1^{n_0}(\mathcal{M}_{n+1}) \subset \mathrm{int}(\mathcal{M}_n)$ for $n \in \mathbb{N}_0$,
[M3] $\lim\limits_{n \to \infty} \mathrm{diam}(\mathcal{M}_n) = 0$,
[M4] the sequence $\{\mathcal{M}_n\}_{n=0}^{\infty}$ has the extension property.

Without loss of generality, we can also assume that $\mathcal{M}_0 \subset B(x_0, \frac{\varepsilon}{3})$. Obviously, there exists an open set $U \subset \mathbb{M}$, such that $x_0 \in U$ and $f_1^{n_0}(U) \subset B(x_0, \frac{\varepsilon}{3})$. Moreover, Condition [M3] implies that there exists $k^* > 1$ such that $\mathcal{M}_k \subset U$ for $k \geq k^*$.

Put $k_1 = k^* + 1$. Since $f_1^{n_0}$ is irreducible at $x_0 \in \mathrm{int}(\mathcal{M}_{k_1})$, it is easy to see that there exist $x_1 \in \mathcal{M}_{k_1}$ and an arc $A(x_0, f_1^{n_0}(x_1))$ with endpoints at x_0 and $f(x_1)$ such that $A(x_0, f_1^{n_0}(x_1)) \subset f_1^{n_0}(\mathcal{M}_{k_1})$. Let A_1^1, A_2^1 be disjoint arcs such that $A_1^1, A_2^1 \subset A(x_0, f_1^{n_0}(x_1))$ and $x_0 \notin A_1^1 \cup A_2^1$. Put $\Gamma_i^1 = f_1^{-n_0}(A_i^1) \cap \mathcal{M}_{k_1}$ for $i = 1, 2$. Then, $\Gamma_1^1 \neq \varnothing \neq \Gamma_2^1$, $x_0 \notin \Gamma_1^1 \cup \Gamma_2^1 \subset \mathcal{M}_{k_1}$, $\Gamma_1^1 \cap \Gamma_2^1 = \varnothing$ and the sets Γ_1^1 and Γ_2^1 are closed. Moreover, $f_1^{n_0}(\Gamma_i^1) = A_i^1$ for $i = 1, 2$.

On account of the well-known Hahn–Mazurkiewicz theorem (see, e.g., [19], p. 106), there exists a continuous function $g_1 : A_1^1 \cup A_2^1 \to \mathcal{M}_{k_1}$ such that $g_1(A_1^1) = \mathcal{M}_{k_1}$ and $g_1(A_2^1) = \mathcal{M}_{k_1}$. From the fact that the set $\Gamma_1^1 \cup \Gamma_2^1$ is closed and from Condition [M3], it follows that there exists $k_2 > k_1$ such that $\mathcal{M}_{k_2} \cap (\Gamma_1^1 \cup \Gamma_2^1) = \varnothing$. Obviously, $(A_1^1 \cup A_2^1) \cap f_1^{n_0}(\mathcal{M}_{k_2}) = \varnothing$.

By the same reasoning as above, one can find $x_2 \in \mathcal{M}_{k_2}$ and $A(x_0, f_1^{n_0}(x_2)) \subset f_1^{n_0}(\mathcal{M}_{k_2})$. Let $A_1^2, A_2^2, A_3^2, A_4^2$ be such arcs that $A_1^2 \cup A_2^2 \cup A_3^2 \cup A_4^2 \subset A(x_0, f_1^{n_0}(x_2))$, $A_i^2 \cap A_j^2 = \varnothing$ if $i \neq j$,

$x_0 \notin A_1^2 \cup A_2^2 \cup A_3^2 \cup A_4^2$. Put $\Gamma_i^2 = f_1^{-n_0}(A_i^2) \cap \mathcal{M}_{k_2}$, $i = 1, \ldots, 4$. Clearly $\bigcup_{i=1}^{4} \Gamma_i^2 \subset \mathcal{M}_{k_2}$, $\Gamma_i^2 \cap \Gamma_j^2 = \varnothing$

whenever $i \neq j$, $x_0 \notin \bigcup_{i=1}^{4} \Gamma_i^2$ and Γ_i^2 are closed for $i = 1, \ldots, 4$. Moreover, $f_1^{n_0}(\Gamma_i^2) = A_i^2$ for $i = 1, \ldots, 4$.

Let $g_2 : A_1^2 \cup A_2^2 \cup A_3^2 \cup A_4^2 \to \mathcal{M}_{k_2}$ be a continuous function such that $g_2(A_i^2) = \mathcal{M}_{k_2}$ for $i = 1, \ldots, 4$.

Continuing in this fashion, we obtain two sequences: $\{k_i\}_{i\in\mathbb{N}} \subset \mathbb{N}$ and $\{\Gamma_i\}_{i\in\mathbb{N}}$ of closed sets such

that $\Gamma_i = \bigcup_{s=1}^{2^i} \Gamma_s^i \subset \mathcal{M}_{k_i}$ for $i \in \mathbb{N}$, Γ_s^i is closed for $i \in \mathbb{N}, s \in \{1, \ldots, 2^i\}$ and $\Gamma_{s_1}^i \cap \Gamma_{s_2}^i = \varnothing$ whenever

$s_1 \neq s_2$. Moreover, there exists a sequence $\{g_i\}_{i\in\mathbb{N}}$ of continuous functions such that $g_i(f_1^{n_0}(\Gamma_s^i)) = \mathcal{M}_{k_i}$ for $i \in \mathbb{N}, s \in \{1, \ldots, 2^i\}$.

Now, let us consider the set $\Gamma = \bigcup_{i=1}^{\infty} \Gamma_i \cup \{x_0\}$. It follows easily that $\Gamma \subset \operatorname{int}(\mathcal{M}_1)$. It is easy to prove that Γ is closed.

Consider the following function:

$$g_0(x) = \begin{cases} x_0 & \text{for } x = x_0, \\ g_i(x) & \text{for } x \in \bigcup_{s=1}^{2^i} \Gamma_s^i, i \in \mathbb{N}, \\ f_1^{n_0}(x) & \text{for } x \in \operatorname{Fr} \mathcal{M}_{k^*}. \end{cases}$$

Clearly, $g_0 : \Gamma \cup \operatorname{Fr} \mathcal{M}_{k^*} \to \mathcal{M}_{k^*-1}$. Since $\Gamma \cup \operatorname{Fr} \mathcal{M}_{k^*}$ is closed and g_0 is continuous, it follows by Condition [M4] that there exists a continuous function $g_0^* : \mathcal{M}_{k^*} \to \mathcal{M}_{k^*-1}$ such that $g_0^* \upharpoonright (\Gamma \cup \operatorname{Fr} \mathcal{M}_{k^*}) = g_0$.

Put:

$$\psi(x) = \begin{cases} g_0^*(x) & \text{for } x \in \mathcal{M}_{k^*}, \\ f_1^{n_0}(x) & \text{for } x \notin \mathcal{M}_{k^*}. \end{cases}$$

Consider the system:

$$T_{1,\infty} = \{f_1, f_2, \ldots, f_{n_0}, \psi, f_1, f_2, \ldots, f_{n_0}, \psi, \ldots\}.$$

We will show that $T_{1,\infty}$ is a periodic ε-disturbance of $f_{1,\infty}$. Condition (PD1) is obvious. To obtain Condition (PD2), it is enough to show that $\psi \in [f_1^{n_0}]_{x_0}^{\varepsilon}$. We have $\neq (\psi, f_1^{n_0}) \subset \mathcal{M}_{k^*} \subset B(x_0, \varepsilon)$ because $\psi(x) = f_1^{n_0}(x)$ for $x \notin \mathcal{M}_{k^*}$. Moreover, $f_1^{n_0}(\neq (\psi, f_1^{n_0})) \subset f_1^{n_0}(\mathcal{M}_{k^*}) \subset f_1^{n_0}(U) \subset B(x_0, \varepsilon)$ and $\psi(\neq (\psi, f_1^{n_0})) \subset \psi(\mathcal{M}_{k^*}) = g_0^*(\mathcal{M}_{k^*}) \subset \mathcal{M}_{k^*-1} \subset B(x_0, \varepsilon)$. This means that $\psi \overset{\varepsilon}{\underset{x_0}{=}} f_1^{n_0}$.

What is left is to prove that x_0 is an asymptotical $\mathfrak{A}$-focal entropy point of $T_{1,\infty}$.

Let V be an arbitrary open neighborhood of x_0. Obviously, there exists $k_0 \in \mathbb{N}$ such that $\mathcal{M}_k \subset V$ for $k > k_0$. Let $\alpha \in \mathbb{R}$ and $\alpha > 0$. We will show that there exists $\mathcal{A} = (A_1, \ldots, A_m) \in \Theta(\mathfrak{A})$ such that $H_{T_{1,\infty}^{n_0+1}}(\mathcal{A}) \geq (n_0 + 1)\alpha$. Obviously, one can find $i^* \in \mathbb{N}$ such that $k_{i^*} > k_0$ and $i^* > (n_0 + 1)\alpha$.

Thus, $\Gamma_{i^*} = \bigcup_{s=1}^{2^{i^*}} \Gamma_s^{i^*} \subset V$. Consider $\mathcal{A} = (\Gamma_1^{i^*}, \ldots, \Gamma_{2^{i^*}}^{i^*}) \subset V$, and put $\tilde{\psi} = \psi \circ f_{n_0} \circ \cdots \circ f_1$. Clearly, $T_{1,\infty}^{n_0+1} = (\tilde{\psi})$.

Let $k \in \mathbb{N}$. It is evident that $\operatorname{tr}(M_{T_{1,\infty}^{n_0+1}}^k(\mathcal{A}))$ is equal to the number of $(k+1)$-paths connected with $\mathcal{A}$. We have $\tilde{\psi}(\Gamma_s^{i^*}) = \psi(f_1^{n_0}(\Gamma_s^{i^*})) = \mathcal{M}_{k_{i^*}}$ and $\Gamma_s^{i^*} \subset \mathcal{M}_{k_{i^*}}$ for any $s \in \{1, \ldots, 2^{i^*}\}$, so $\Gamma_{s_1}^{i^*} \xrightarrow{\tilde{\psi}} \Gamma_{s_2}^{i^*}$

for $s_1, s_2 \in \{1, \ldots, 2^{i^*}\}$. As a consequence $\operatorname{tr}(M_{T_{1,\infty}^{n_0+1}}^k(\mathcal{A})) = (2^{i^*})^k$. Thus, $H_{T_{1,\infty}^{n_0+1}}(\mathcal{A}) = \log 2^{i^*} = i^* > (n_0 + 1)\alpha$.

Finally, we have shown that for any $\alpha > 0$, there exists $\mathcal{A} \in \Theta(\mathfrak{A})$, $\mathcal{A} \subset V$ such that $\frac{1}{n_0+1} H_{T_{1,\infty}^{n_0+1}}(\mathcal{A}) > \alpha$. Hence for any $\alpha > 0$, we have $H(\mathfrak{A}, T_{1,\infty}, V) = \sup\{\frac{1}{n} H_{T_{1,\infty}^n}(\mathcal{A}) : \mathcal{A} \in \Theta(\mathfrak{A}) \wedge$

$\mathcal{A} \subset V \wedge n \in \mathbb{N}\} \geq \sup\{\frac{1}{n_0+1} H_{T_{1,\infty}^{n_0+1}}(\mathcal{A}) : \mathcal{A} \in \Theta(\mathfrak{A}) \wedge \mathcal{A} \subset V\} \geq \alpha$. Thus, $H(\mathfrak{A}, T_{1,\infty}, V) = +\infty$, and therefore, $d^*(\mathcal{A}, T_{1,\infty}, V) = 1$, so x_0 is an asymptotical $\mathfrak{A}$-focal entropy point of $T_{1,\infty}$. $\quad\square$

The next theorem shows the difference between a $\mathfrak{A}$-focal entropy point of NDS and an asymptotical $\mathfrak{A}$-focal entropy point of NDS on the interval under as weak as possible assumptions imposed on the considered functions. For the simplicity of the notation, we will formulate and prove the theorem for $x_0 = 0$. It can be easily generalized for any $x_0 \in [0,1]$.

Theorem 4. *Let $f : [0,1] \to [0,1]$ be a function continuous at $0 \in \text{FIX}(f)$ and such that $h(f) < \infty$. Let us assume that:*

(*) *there exists a sequence $\alpha_n \searrow 0$ such that for any $n \in \mathbb{N}$, we have $f([\alpha_n, 1]) \subset [\alpha_n, 1]$.*

Then, for any $\varepsilon > 0$, there exists a sequence $\{f_n\}_{n\in\mathbb{N}}$ of functions continuous at zero such that $\{f_n\}_{n\in\mathbb{N}} \subset [f]_{x_0}^\varepsilon$ and zero is a $\mathfrak{A}$-focal entropy point of the system $f_{1,\infty}$ and is not an asymptotical $\mathfrak{A}$-focal entropy point of $f_{1,\infty}$.

Proof. Let $\varepsilon > 0$. Let γ be a positive number less than ε and such that $f(x) < \varepsilon$ for $x \in [0,\gamma]$. There exists $n_0 \in \mathbb{N}$ such that $\alpha_{n_0} \in (0,\gamma)$ and $f([\alpha_{n_0}, 1]) \subset [\alpha_{n_0}, 1]$. Put $\delta = \alpha_{n_0}$, and hence, $f(\delta) \geq \delta$. Let $m \in \mathbb{N}$ be an odd positive integer such that $\log m > h(f)$.

From (*), it follows that there exists an interval $P \subset (0,\delta)$ such that $f([\delta, 1]) \cap P = \varnothing$. Put $a_0 = \inf P$ and $b_0 = \sup P$. Notice that $0 < b_0 < \delta$. Consider a sequence $x_n \searrow 0$ such that $x_1 = a_0$. Now, we can define the function $f_1 : [0,1] \to [0,1]$ as follows: $f_1(0) = 0$, $f_1(x_{n+1} + 2k\frac{x_n - x_{n+1}}{m}) = x_{n+1}$ for $k \in \{0, 1, \dots, \frac{m-1}{2}\}$, $f_1(x_{n+1} + (2k-1)\frac{x_n - x_{n+1}}{m}) = x_n$ for $k \in \{1, \dots, \frac{m+1}{2}\}$, $f_1(x_{n+1} + \frac{x_n - x_{n+1}}{2m}) = x_{n+1} + \frac{x_n - x_{n+1}}{2m}$, $f_1(x_n - \frac{x_n - x_{n+1}}{2m}) = x_n - \frac{x_n - x_{n+1}}{2m}$, f_1 is linear on respective intervals in each $[x_{n+1}, x_n]$; and moreover, $f_1(x) = a_0$ for $x \in [a_0, b_0)$, $f_1(x) = b_0$ for $x \in [b_0, \delta)$ and $f_1(x) = f(x)$ for $x \in [\delta, 1]$.

We next define functions f_n for $n \geq 2$. Let $f_n(x) = f_1(x)$ for $x \in [0,1] \setminus (a_0, \delta)$, $n \geq 2$. Fix $y_0 \in (a_0, b_0)$. Put $f_n(a_0 + 2k\frac{y_0 - a_0}{m+2}) = a_0$ for $k \in \{0, 1, \dots, \frac{m+1}{2}\}$ and $f_n(a_0 + (2k-1)\frac{y_0 - a_0}{m+2}) = y_0$ for $k \in \{1, \dots, \frac{m+3}{2}\}$ and f_n linear on the respective intervals. Moreover, $f_n(x) = y_0$ for $x \in [y_0, b_0)$, $f_n(x) = b_0$ for $x \in [b_0, \delta)$.

Obviously f_n is continuous at zero for $n \in \mathbb{N}$ and $\{f_n\}_{n\in\mathbb{N}} \subset [f]_0^\varepsilon$. We will show that $h(f_1) = \log m = h(f_{1,\infty})$ and $h(f_n) = \log(m+2) = h(f_{n,\infty})$ for $n \geq 2$.

We first prove that $h(f_1, [a_0, b_0)) = 0$. Let $\varepsilon_1 > 0$, $n \in \mathbb{N}$ and $M \subset [a_0, b_0)$ be an (n, ε_1)-separated set for f_1. For any $x, y \in M$, $x \neq y$, there exists $i_0 \in \{0, \dots, n-1\}$ such that $\rho((f_1)^{i_0}(x), (f_1)^{i_0}(y)) > \varepsilon_1$. Notice that $i_0 = 0$. Indeed, we have $f_1(x) = a_0$ and $f_1(y) = a_0$. Hence, for $i > 0$, we have $(f_1)^i(x) = a_0$ and $(f_1)^i(y) = a_0$, so $\rho((f_1)^i(x), (f_1)^i(y)) = 0$ for $i > 0$. As a consequence, for any distinct points $x, y \in M$, we have $\rho(x,y) > \varepsilon_1$. It follows that $\#(M) \leq [\frac{b_0 - a_0}{\varepsilon_1}] + 1$, so $s_n(f_1, [a_0, b_0), \varepsilon_1) \leq [\frac{b_0 - a_0}{\varepsilon_1}] + 1$, where $[\frac{b_0 - a_0}{\varepsilon_1}]$ denotes the smallest positive integer greater than $\frac{b_0 - a_0}{\varepsilon_1}$. Hence, $h(f_1, [a_0, b_0)) = \lim\limits_{\varepsilon_1 \to 0} \limsup\limits_{k \to \infty} \frac{1}{k} \log(s_k(f_1, [a_0, b_0), \varepsilon_1)) \leq 0$. In an analogous way, one can show that $h(f_1, [b_0, \delta)) = 0$.

Moreover, we have $f_1(x) = f(x)$ for $x \in [\delta, 1]$ and $f_1([\delta, 1]) \subset [\delta, 1]$. Consequently, $h(f_1, [\delta, 1]) = h(f, [\delta, 1]) < \log m$.

Let $n \in \mathbb{N}$. We will show that $h(f_1, [x_{n+1}, x_n]) = \log m$. Clearly, $f_1 \upharpoonright [x_{n+1}, x_n] : [x_{n+1}, x_n] \to [x_{n+1}, x_n]$ and $f_1 \upharpoonright [x_{n+1}, x_n]$ is piecewise monotone. Denote by c_k the number of intervals of monotonicity of $(f_1)^k$. We have $c_k = m^k$ for $k \in \mathbb{N}$. Thus, by Theorem 4.2.4 [20], we have $h(f_1, [x_{n+1}, x_n]) = \lim\limits_{k\to\infty} \frac{1}{k} \log c_k = \log m$. Obviously, $[0, a_0] = \bigcup\limits_{n\in\mathbb{N}} [x_{n+1}, x_n]$, and for any $n \in \mathbb{N}$, we have $f_1([x_{n+1}, x_n]) \subset [x_{n+1}, x_n]$. On account of Lemma 4.1.10 [20] (and the remark after it), we obtain $h(f_1, [0, a_0]) = \sup\limits_{n\in\mathbb{N}}(h(f_1, [x_{n+1}, x_n])) = \log m$. Finally, Proposition 3.5 [12] (see also Lemma 4.1 from [13]) gives that $h(f_1) = \max\{h(f_1, [0, a_0]), h(f_1, [a_0, b_0)), h(f_1, [b_0, \delta)), h(f_1, [\delta, 1])\} = \log m$.

We now turn to the case $n \geq 2$. We have $f_n \upharpoonright [0, a_0] = f_1 \upharpoonright [0, a_0]$ and $f_1 : [0, a_0] \to [0, a_0]$. Hence, $h(f_n, [0, a_0]) = h(f_1, [0, a_0]) = \log m$. Moreover, $f_n \upharpoonright [\delta, 1] = f_1 \upharpoonright [\delta, 1]$ and $f_1 : [\delta, 1] \to [\delta, 1]$, so $h(f_n, [\delta, 1]) = h(f_1, [\delta, 1]) < \log m$. Therefore, $f_n \upharpoonright [b_0, \delta) = f_1 \upharpoonright [b_0, \delta)$ and $f_1 : [b_0, \delta) \to [b_0, \delta)$, so $h(f_n, [b_0, \delta)) = h(f_1, [b_0, \delta)) = 0$.

As was done for a function f_1, one can show that $h(f_n, [y_0, b_0)) = 0$ and $h(f_n, [a_0, y_0]) = \log(m + 2)$. As a consequence, by Proposition 3.5 [12], we obtain $h(f_n) = \log(m + 2)$.

Since $f_n = f_2$ for $n \geq 2$, it follows that $h(f_{n,\infty}) = h(f_2) = \log(m + 2)$ for $n \geq 2$.

We will show now that $h(f_{1,\infty}) = \log m$. We claim that:

$$(f_1)^i(z) = f_1^i(z) \text{ for } z \in [0, 1] \text{ and } i \in \mathbb{N}_0. \tag{10}$$

Indeed, if $z \in (a_0, b_0)$, then $f_1(z) = a_0$. Thus, for $i \geq 1$, we have $(f_1)^i(z) = a_0$ and $f_1^i(z) = a_0$, so $(f_1)^i(z) = f_1^i(z)$. For $i = 0$, we have $(f_1)^0(z) = z = f_1^0(z)$, so $(f_1)^i(z) = f_1^i(z)$ for $z \in (a_0, b_0)$ and $i \in \mathbb{N}_0$. If $z \in [0, 1] \setminus (a_0, b_0)$ then $f_n(z) \in [0, 1] \setminus (a_0, b_0)$ for $n \in \mathbb{N}$. Therefore, it is easy to see that for $z \in [0, 1] \setminus (a_0, b_0)$ and $i \geq 1$, we have $(f_1)^i(z) = f_1^i(z)$. Obviously, $(f_1)^0(z) = z = f_1^0(z)$. The proof of (10) is complete.

Notice that for any $n \in \mathbb{N}$ and $\varepsilon_1 > 0$, the set $M \subset [0, 1]$ is (n, ε_1)-separated for the system $f_{1,\infty}$ if and only if M is (n, ε_1)-separated for f_1. Indeed, let M be an (n, ε_1)-separated set for f_1. Then, for any distinct points $x, y \in M$, there exists $i \in \{0, \ldots, n - 1\}$ such that $\rho((f_1)^i(x), (f_1)^i(y)) > \varepsilon_1$. By (10), we obtain $\rho(f_1^i(x), f_1^i(y)) > \varepsilon_1$, which means that M is an (n, ε_1)-separated set for the system $f_{1,\infty}$. The proof of the converse implication runs in a similar way.

As a consequence, we have $s_n(f_1, [0, 1], \varepsilon_1) = s_n(f_{1,\infty}, [0, 1], \varepsilon_1)$, so $\log m = h(f_1) = h(f_{1,\infty})$.

Let U be an arbitrary neighborhood of zero. We will show that $H(\mathfrak{A}, f_{1,\infty}, U) = \log m$. Clearly, by Theorem 1, we have:

$$H(\mathfrak{A}, f_{1,\infty}, U) \leq h(f_{1,\infty}) = \log m. \tag{11}$$

Let $n \in \mathbb{N}$. Consider the interval $[x_{n+1}, x_n]$. There exists a sequence of points $x_{n+1} < a_{n,1} < b_{n,1} < a_{n,2} < b_{n,2} < \cdots < a_{n,m} < b_{n,m} < x_n$ such that $f_1([a_{n,i}, b_{n,i}]) = [a_{n,i}, b_{n,i}]$ for $i \in \{1, \ldots, m\}$. Put $A_i^n = [a_{n,i}, b_{n,i}]$ for $i \in \{1, \ldots, m\}$. Then, $A^n = (A_1^n, \ldots, A_m^n) \in \Theta(\mathfrak{A})$ and:

$$\text{for any } k \in \mathbb{N} \text{ and any } i, j \in \{1, \ldots, m\} \text{ we have } A_i^n \underset{f_k}{\to} A_j^n. \tag{12}$$

Obviously, for any $k \in \mathbb{N}$, the trace $\text{tr}(M_{f_{1,\infty}}^k(A^n))$ is equal to the number of $(k + 1)$-paths with the first and the last node at A_i^n for $i = 1, \ldots, m$. By (12), we conclude that the number of such paths is equal to m^k. Hence, $\frac{1}{k} \log \text{tr}(M_{f_{1,\infty}}^k(A^n)) = \log m$, and therefore:

$$H_{f_{1,\infty}}(A^n) = \log m. \tag{13}$$

Let $n_0 \in \mathbb{N}$ be such that $[x_{n_0+1}, x_{n_0}] \subset U$. Then, by (13), we obtain $H_{f_{1,\infty}}(A^{n_0}) = \log m$, so $H(\mathfrak{A}, f_{1,\infty}, U) \geq \log m$. From this and (11), we get $H(\mathfrak{A}, f_{1,\infty}, U) = \log m$. As a consequence, $d(\mathfrak{A}, f_{1,\infty}, U) = 1$, which gives $E(\mathfrak{A}, f_{1,\infty}, 0) = 1$, so zero is a $\mathfrak{A}$-focal entropy point of $f_{1,\infty}$.

Simultaneously, zero is not an asymptotical $\mathfrak{A}$-focal entropy point of $f_{1,\infty}$, because for any neighborhood U of zero, we have $H(\mathfrak{A}, f_{1,\infty}, U) = \log m$ and $h^*(f_{1,\infty}) = \lim_{n \to \infty} h(f_{n,\infty}) = \log(m + 2)$. Therefore, $d^*(\mathfrak{A}, f_{1,\infty}, U) = \frac{\log m}{\log(m+2)}$, which means that $E^*(\mathfrak{A}, f_{1,\infty}, 0) = \frac{\log m}{\log(m+2)} < 1$. $\square$

4. Conclusions

In the paper, the notions of a focal entropy point and an asymptotical focal entropy point for nonautonomous dynamical systems are introduced. The definitions adopted in the paper specify the notions that express the complexity of a system around these points and moreover, the complexity of a system around such points does not depend on the behavior of the system in other parts of its domain. Each asymptotical focal entropy point of an NDS is its focal entropy point. In the case of periodic dynamical systems these notions coincide. For a periodic NDS consisting of continuous functions defined on the closed unit interval there exists an asymptotical focal entropy point. Moreover, there exists a dynamical system with a focal entropy point which is not its asymptotical focal entropy point. In the case of some periodic dynamical systems consisting of continuous functions defined on a topological manifold one can disturb a system to obtain a system "lying close" to the given one and having an asymptotical focal entropy point.

Author Contributions: The authors contributed equally to this work. All the authors took part in all actions connected with this work.

Conflicts of Interest: The authors declare no conflict of interest.

References

1. Alsedá, L.; Cushing, J.M.; Elaydi, S.; Pinto, A.A. Difference Equationes, Discrete Dynamical Systems and Applications. In *Proceedings in Mathematics and Statistics*; Springer: Berlin, Germany, 2016.
2. Cánovas, J.S.; Muñoz-Guillermo M. On the complexity of economic dynamics: An approach through topological entropy. *Chaos Soliton. Fract.* **2017**, *103*, 163–176.
3. Elaydi, S.; Sacker, R.J. Global stability of periodic orbits of nonautonomous difference equations in population biology. *J. Differ. Equ.* **2005**, *208*, 258–273.
4. François, M.; Grosges, T.; Barchiesi, D.; Erra, R. Pseudo-random number generator based on mixing of three chaotic maps. *Commun. Nonlinear Sci. Numer. Simulat.* **2014**, *19*, 887–895.
5. Gandomi, A.H.; Yun, G.J.; Yang, X.S.; Talatahari S. Chaos-enhanced accelerated particle swarm optimization. *Commun. Nonlinear Sci. Numer. Simulat.* **2013**, *18*, 327–340.
6. Luis, R.; Elaydi S.; Oliveira, H. Nonautonomous periodic systems with Allee efects. *J. Difference Equ. Appl.* **2010**, *16*, 1179–1196.
7. Yakubu, A.A.; Castillo-Chavez, C. Interplay between local dynamics and dispersal in discrete-time metapopulation models. *J. Theoret. Biol.* **2002**, *218*, 273–288.
8. Pawlak, R.J. *Entropy of Nonautonomous Discrete Dynamical Systems Considered in GTS and GMS*; Bulletin de la Société des Sciences et des Lettres de Łódź LXVI. Lodzkie Towarzystwo Naukowe: Łódź, Poland, 2016.
9. Adler, R.L.; Konheim, A.G.; McAndrew, M.H. Topological entropy. *Trans. Amer. Math. Soc.* **1965**, *114*, 309–319.
10. Bowen, R. Entropy for group endomorphisms and homogeneous spaces. *Trans. Amer. Math. Soc.* **1971**, *153*, 401–414; Erratum in **1973**, *181*, 509–510.
11. Dinaburg, E.I. Connection between various entropy characterizations of dynamical systems. *Izv. Akad. Nauk SSSR Ser. Mat.* **1971**, *35*, 324–366. (In Russian).
12. Čiklová, M. Dynamical systems generated by functions with G_δ graphs. *Real Anal. Exch.* **2004**, *30*, 617–638.
13. Kolyada, S.; Snoha, L. Topological entropy of nonautonomous dynamical systems. In *Random & Computational Dynamics*; Marcel Dekker: New York, NY, USA, 1996.
14. Nitecki, Z.H. Topological entropy and the preimage structure of maps. *Real Anal. Exch.* **2003**, *29*, 9–42.
15. Ye, X.; Zhang, G. Entropy points and applications. *Trans. Amer. Math. Soc.* **2007**, *359*, 6167–6186.
16. Korczak-Kubiak, E.; Loranty, A.; Pawlak, R.J. On Focusing Entropy at a Point. *Taiwanese J. Math.* **2016**, *20*, 1117–1137.
17. Kawan, C. Metric entropy of nonautonomous dynamical system. *Nonauton. Stoch. Dyn. Syst.* **2013**, *1*, 26–52.
18. Lee, J.M. *Introduction to Topological Manifolds*; Springer: Berlin, Germany, 2000.
19. Sagan, H. *Space-Filling Curves*; Springer: Berlin, Germany, 1994.
20. Alsedá, L.; Llibre, J.; Misiurewicz, M. *Combinatorial Dynamics and Entropy in Dimension One*; World Scientific: Singapore, 1993.

Article

Analytical Solutions for Multi-Time Scale Fractional Stochastic Differential Equations Driven by Fractional Brownian Motion and Their Applications

Xiao-Li Ding [1,*] and Juan J. Nieto [2]

1   Department of Mathematics, Xi'an Polytechnic University, Xi'an 710048, China
2   Departamento de Estatística, Análisis Matemático y Optimización, Universidad de Santiago de Compostela, 15782 Santiago de Compostela, Spain; juanjose.nieto.roig@usc.es
*   Correspondence: dingding0605@126.com; Tel.: +86-158-2968-0246

Received: 21 December 2017; Accepted: 11 January 2018; Published: 16 January 2018

Abstract: In this paper, we investigate analytical solutions of multi-time scale fractional stochastic differential equations driven by fractional Brownian motions. We firstly decompose homogeneous multi-time scale fractional stochastic differential equations driven by fractional Brownian motions into independent differential subequations, and give their analytical solutions. Then, we use the variation of constant parameters to obtain the solutions of nonhomogeneous multi-time scale fractional stochastic differential equations driven by fractional Brownian motions. Finally, we give three examples to demonstrate the applicability of our obtained results.

Keywords: multi-time scale fractional stochastic differential equations; fractional Brownian motion; fractional stochastic partial differential equation; analytical solution

---

1. Introduction

In the last few years, the interest of the scientific community towards fractional calculus has experienced an exceptional boost, and so its applications can now be found in a great variety of scientific fields—for example, anomalous diffusion [1–3], medicine [4], solute transport [5], random and disordered media [6–8], information theory [9], electrical circuits [10], and so on. The reason for the success of fractional calculus in modeling natural phenomena is that the operators are nonlocal, which makes them suitable to describe the long memory or nonlocal effects characterizing most physical phenomena.

Fractional stochastic differential equations (FSDEs) are an important class of differential equations. They can model the dynamics of complex systems in finance [8,11–13], and in physical problems [14,15]. For example, in [8], the authors combined stochastic contact process and compound Poisson process to construct a novel microscope complex price dynamics, in an attempt to reproduce and characterize the complex dynamics of financial markets. In finance, fractional permutation entropy, sample entropy, and fractional sample entropy play important roles. It is well-known that entropy is used to quantify the complexity and uncertainty in financial time series and others. At the same time, the necessity of a powerful technique for solving these new types of equations arose, becoming one of the main research objects in the fields of theoretical and applied sciences. In the available literature, there exist various methods for solving fractional stochastic differential equations, such as analytical methods and numerical algorithms [16–31].

Analytical solutions of fractional partial equations are of fundamental importance in describing and understanding physical phenomena, since all the parameters are expressed in the form of infinite series, and therefore the influence of individual parameters on natural phenomena can be easily examined. Additionally, the analytical solutions make it easy to study asymptotic behaviors of the solutions, which are usually difficult to obtain through numerical calculations. Besides, the analytical solutions may serve as tools in assessing the computational performance and accuracy of numerical solutions. Especially, for stochastic differential systems, analytical solutions may provide a useful tool for assessing the influence of some parameters on statistical properties, permutation entropy, fractional permutation entropy, sample entropy, and fractional sample entropy. It is well-known that entropy theory is an important issue because it enables hydraulic and control engineers to quantify uncertainties, determine risk and reliability, estimate parameters, model processes, and design more robust and reliable hydraulic canals control systems.

To the authors' knowledge, the analytical solutions of the FSDEs driven by fractional Brownian motions (fBms) have not yet been reported in the literature. X.J. Wang et al. [32] considered the following semilinear parabolic SPDEs in V, driven by an infinite dimensional fractional Brownian motion,

$$dX(t) + AX(t)dt = F(X(t))dt + \Phi dB_H(t), X(0) = x_0, t \in [0, T], \tag{1}$$

where $F : V \to V$ and $\Phi : V \to V$ are deterministic mappings. Motivated by their work, we investigate the analytical solution of the following multi-time scale fractional stochastic differential equation:

$$\frac{dY(t)}{dt} + D_t^\alpha \big(a(t)Y(t) + p(t)\big) = \big(b(t)Y(t) + q(t)\big) + \big(\sigma(t)Y(t) + v(t)\big)\frac{dB_H(t)}{dt}, Y(0) = y_0, \tag{2}$$

where $b, p, \sigma, q, a, v \in C([0, T]), 0 < \alpha \leq 1, B_H$ is a fractional Brownian motion defined on $[0, T]$, and y_0 is a real-valued random variable on a complete probability space $(\Omega, \mathcal{F}, \mathcal{P})$, and it is independent of $B_H(t)$ for all $t \in [0, T]$. The detailed definitions of the Riemann–Liouville fractional derivative and the fractional Laplacian operator and fBm are given in the next section (or see [33–37]).

The rest of this paper is organized as follows. In Section 2, we give some basic definitions of fBm and fractional calculus, which will be used in the paper. In Section 3, we give the solution of multi-time scale FSDEs driven by fBms. In Section 4, we give three examples to demonstrate the applicability of the obtained results. In Section 5, we give conclusions.

2. Preliminaries

In this section, we give some basic definitions of fractional Brownian motion and fractional calculus, which will be used throughout this paper. For details, one can refer to [37–40].

Let $(\Omega, \mathcal{F}, \mathcal{P})$ be a complete probability space, and $[0, T]$ be a finite time interval.

Definition 1. *A one-dimensional fractional Brownian motion $B_H = \{B_H(t), t \in [0, T]\}$ of Hurst index $H \in (0, 1)$ on $[0, T]$ is a continuous and centered Gaussian process on some probability space $(\Omega, \mathcal{F}, \mathcal{P})$ with covariance function*

$$E[B_H(t)B_H(s)] = \frac{1}{2}(t^{2H} + s^{2H} - |t - s|^{2H}), \quad t, s \in [0, T].$$

If $H = \frac{1}{2}$, then the corresponding fBm is the usual standard Brownian motion. If $H > \frac{1}{2}$, then the process fBm exhibits a long-range dependence. In this paper, we always assume $H \in (\frac{1}{2}, 1)$.

Lemma 1. *(Fractional Itô formula) [39] If $X(t)$ satisfies that*

$$dX(t) = u(t)dt + v(t)dB_H(t), \tag{3}$$

where u, v are given functions. Furthermore, let $f \in C^2(\mathbb{R})$, and assume that $f'(X)$ and $f''(X)$ exist and are continuous for $X \in \mathbb{R}$. Then, it has

$$df(X(t)) = (f'(X(t))u(t) + Hf''(X(t))t^{2H-1}v^2(t))dt + f'(X(t))v(t)dB_H(t). \tag{4}$$

It is interesting to note that if $H = \frac{1}{2}$ is formally substituted in Equation (4), then the well-known Itô formula for classical Brownian motion is obtained.

In the following, we recall some definitions about fractional calculus and some special functions.

Definition 2. *Let $\alpha > 0$. Then, the Riemann–Liouville fractional integral of order α with respect to t is defined as*

$$\mathcal{I}_t^\alpha f(t) = \frac{1}{\Gamma(\alpha)} \int_0^t (t - \tau)^{\alpha - 1} f(\tau) d\tau, \quad t > 0, \tag{5}$$

where $\Gamma(\cdot)$ is the Gamma function.

Definition 3. *Let $f \in C([0, T])$ and $m - 1 < \alpha \leq m$, where $m \in \mathbb{N}^+$. The Riemann–Liouville fractional derivative of order α with respect to t is defined as*

$$D_t^\alpha f(t) = \frac{1}{\Gamma(m - \alpha)} \frac{d^m}{dt^m} \int_0^t (t - \tau)^{m - \alpha - 1} f(\tau) d\tau, \quad t > 0. \tag{6}$$

There exists the following relationship between the Riemann–Liouville fractional integral and the Riemann–Liouville fractional derivative.

Property 1. *Let $m - 1 < \alpha \leq m$, where $m \in \mathbb{N}^+$ [37]. Then the statements are true:*

$$(D_t^\alpha \mathcal{I}_t^\alpha f)(t) = f(t), \quad (\mathcal{I}_t^\alpha D_t^\alpha f)(t) = f(t) - \sum_{k=1}^m \frac{(\mathcal{I}_t^{m-\alpha} f)^{(m-k)}(0^+)}{\Gamma(\alpha - k + 1)} t^{\alpha - k}, \quad t > 0. \tag{7}$$

Definition 4. *Suppose that the Laplacian $(-\triangle)$ has a complete set of orthonormal eigenfunctions φ_n corresponding to eigenvalues λ_n^2 on a bounded region D; i.e., $(-\triangle)\varphi_n = \lambda_n^2 \varphi_n$ on D; $\mathcal{B}(\varphi_n) = 0$ on ∂D, where $\mathcal{B}(\varphi_n)$ is one of the standard three homogeneous boundary conditions [33]. Let*

$$\mathcal{G} = \left\{ g = \sum_{n=1}^\infty c_n \varphi_n, \ c_n = \langle g, \varphi_n \rangle, \ \sum_{n=1}^\infty |c_n|^2 |\lambda_n|^\alpha < \infty \right\}, \tag{8}$$

then for any $g \in \mathcal{G}$, $(-\triangle)^{\frac{\alpha}{2}}$ is defined by

$$(-\triangle)^{\frac{\alpha}{2}} g = \sum_{n=1}^\infty c_n \lambda_n^\alpha \varphi_n. \tag{9}$$

Lemma 2. *Suppose that the one-dimensional Laplacian $(-\triangle)$ defined with Dirichlet boundary conditions at $x = 0$ and $x = L$ has a complete set of orthonormal eigenfunctions φ_n corresponding to eigenvalues λ_n^2 on a bounded region $[0, L]$ [33]. If $(-\triangle)\varphi_n = \lambda_n^2 \varphi_n$ on $[0, L]$, and $\varphi_n(0) = \varphi_n(L) = 0$, then, the eigenvalues are given by $\lambda_n^2 = \frac{n^2 \pi^2}{L^2}$, and the corresponding eigenfunctions are $\varphi_n(x) = \sin(n\pi x / L)$, $n = 1, 2, \ldots$.*

Definition 5. *The two-parameter Mittag–Leffler function is defined by [37]*

$$E_{\alpha, \beta}(z) = \sum_{k=0}^\infty \frac{z^k}{\Gamma(\alpha k + \beta)}, \quad \alpha, \beta > 0. \tag{10}$$

The one-parameter Mittag–Leffler function is defined by

$$E_\alpha(z) = \sum_{k=0}^\infty \frac{z^k}{\Gamma(\alpha k + 1)}, \quad \alpha > 0. \tag{11}$$

In particular, when $\beta = 1$, the two-parameter Mittag–Leffler function coincides with the one-parameter Mittag–Leffler function; i.e., $E_{\alpha,1}(z) = E_\alpha(z)$.

Definition 6. *A generalized Mittag–Leffler function is defined by* [37]

$$E_{\alpha,m,l}(z) = \sum_{k=0}^{\infty} c_k z^k, \tag{12}$$

with

$$c_0 = 1, \quad c_k = \prod_{j=0}^{k-1} \frac{\Gamma(\alpha(jm+l)+1)}{\Gamma(\alpha(jm+l+1)+1)}, \tag{13}$$

where $\alpha > 0$, $m > 0$, and $\alpha(jm+l) > 0$.

In particular, when $m = 1$, there exists the following relationship between the generalized Mittag–Leffler function and the two-parameter Mittag–Leffler function:

$$E_{\alpha,1,l}(z) = \Gamma(\alpha l + 1)E_{\alpha,\alpha l+1}(z). \tag{14}$$

3. Solution Representation for FSDEs Driven by fBms

In this section, we first give an equivalent form of Equation (2) and then investigate its analytical solution. Before giving its equivalent form, we provide some explanations about the Riemann–Liouville fractional integral. In [29] (Definition 3.2 and Example 3.1), the authors gave that the integral with respect to $(dt)^\alpha$ defined as

$$\int_0^t f(\tau)(d\tau)^\alpha = \alpha \int_0^t (t-\tau)^{\alpha-1} f(\tau)d\tau, \quad t > 0, \tag{15}$$

where $f \in C([0,T])$ and $0 < \alpha \le 1$. Based on this definition, we can obtain the following relationship between the Riemann–Liouville fractional integral and the integral with respect to $(dt)^\alpha$:

$$\int_0^t f(\tau)(d\tau)^\alpha = \alpha\Gamma(\alpha)(\mathcal{I}_t^\alpha f)(t). \quad t > 0, \tag{16}$$

where $f \in C([0,T])$ and $0 < \alpha \le 1$.

One sees that Equation (2) is equivalent to the following integral equation:

$$\begin{aligned}
Y(t) \;=\; & y_0 - \frac{1}{\Gamma(1-\alpha)} \int_0^t (t-\tau)^{-\alpha}\big(a(\tau)Y(\tau)+p(\tau)\big)d\tau + \int_0^t \big(b(\tau)Y(\tau)+q(\tau)\big)d\tau \\
& + \int_0^t \big(\sigma(\tau)Y(\tau)+v(\tau)\big)dB_H(\tau).
\end{aligned} \tag{17}$$

By (15), the above equation can be rewritten as:

$$\begin{aligned}
Y(t) \;=\; & y_0 - \frac{1}{\Gamma(2-\alpha)} \int_0^t \big(a(\tau)Y(\tau)+p(\tau)\big)(d\tau)^{1-\alpha} + \int_0^t \big(b(\tau)Y(\tau)+q(\tau)\big)d\tau \\
& + \int_0^t \big(\sigma(\tau)Y(\tau)+v(\tau)\big)dB_H(\tau).
\end{aligned} \tag{18}$$

That is to say, Equation (2) is equivalent to the following equation:

$$\begin{cases} dY(t) = \frac{1}{\Gamma(2-\alpha)}\big(a(t)Y(t)+p(t)\big)(dt)^{1-\alpha} + \big(b(t)Y(t)+q(t)\big)dt + \big(\sigma(t)Y(t)+v(t)\big)dB_H(t), \\ Y(0) = y_0. \end{cases} \tag{19}$$

Therefore, we only need to solve Equation (19). For obtaining the solution of Equation (19), we first discuss the solution of the correspondent homogeneous case in the next subsection.

3.1. Solution Representation for Linear Homogeneous Case

The corresponding homogeneous differential equation can be written as:

$$\begin{cases} dY(t) = \frac{1}{\Gamma(2-\alpha)}a(t)Y(t)(dt)^{1-\alpha} + b(t)Y(t)dt + \sigma(t)Y(t)dB_H(t), \\ Y(0) = y_0. \end{cases} \tag{20}$$

To obtain the solution of Equation (20), we decompose Equation (20) into three subequations:

$$dY^f(t) = \frac{1}{\Gamma(2-\alpha)}a(t)Y^f(t)(dt)^{1-\alpha}, \quad Y^f(0) = y_0^f, \tag{21}$$

$$dY^d(t) = b(t)Y^d(t)dt, \quad Y^d(0) = y_0^d, \tag{22}$$

$$dY^s(t) = \sigma(t)Y^s(t)dB_H(t), \quad Y^s(0) = y_0^s, \tag{23}$$

where y_0^f, y_0^d, y_0^s are constants which satisfy $y_0^f y_0^d y_0^s = y_0$. Obviously, we have

$$\begin{aligned} d(Y^f Y^d Y^s) &= Y^d Y^s(dY^f) + Y^f Y^s(dY^d) + Y^f Y^d(dY^s) \tag{24} \\ &= Y^d Y^s \frac{1}{\Gamma(2-\alpha)}a(t)Y^f(t)(dt)^{1-\alpha} + Y^f Y^s b(t)Y^d(t)dt + Y^f Y^d \sigma(t)Y^s(t)dB_H(t) \tag{25} \\ &= \frac{1}{\Gamma(2-\alpha)}a(t)Y(t)(dt)^{1-\alpha} + b(t)Y(t)dt + \sigma(t)Y(t)dB_H(t). \tag{26} \end{aligned}$$

This implies that $Y = Y^f Y^d Y^s$ is the solution of Equation (20).

In the following, our aim is to solve Equations (21) and (23), because the solution of Equation (22) is well-known. Firstly, we consider the solution of Equation (21).

Lemma 3. *Let $0 < \alpha < 1$ and $a \in C([0, T])$. Then, the solution of Equation (21) is given by*

$$Y^f(t) = \sum_{i=0}^{\infty} \mathcal{R}_a^i y_0^f, \tag{27}$$

where $\mathcal{R}_a$ is an operator defined on $C([0, T])$:

$$(\mathcal{R}_a \varphi)(t) = \frac{1}{\Gamma(1-\alpha)} \int_0^t (t-\tau)^{-\alpha} a(\tau)\varphi(\tau)d\tau, \tag{28}$$

and $\mathcal{R}_a^0$ is an identity operator, and $\mathcal{R}_a^i$ denotes the i-times composition operator of $\mathcal{R}_a$, $i = 1, 2, \dots$.

Proof. Note that Equation (21) is equivalent to the following integral equation:

$$Y(t) = y_0 + \frac{1}{\Gamma(1-\alpha)} \int_0^t (t-\tau)^{-\alpha} a(\tau)Y(\tau)d\tau. \tag{29}$$

Construct a successive approximate sequence $\{Y_{(k)}^f\}$ defined as:

$$Y_{(k+1)}^f(t) = y_0 + \frac{1}{\Gamma(1-\alpha)} \int_0^t (t-\tau)^{-\alpha} a(\tau)Y_{(k)}^f(\tau)d\tau, \quad k = 0, 1, 2, \dots, \tag{30}$$

where we choose $Y^f_{(0)}(t) \equiv y^f_0$. Then, by induction on k, we can obtain

$$Y^f_{(k)}(t) = \sum_{i=0}^{k} \mathcal{R}^i_a y^f_0, \quad k = 0, 1, 2, \ldots, \tag{31}$$

where the operator $\mathcal{R}_a$ is defined in (28).

Next, we will show that the series $\sum\limits_{i=0}^{\infty} (\mathcal{R}^i_a y^f_0)(t)$ is uniformly convergent with respect to $t \in [0, T]$. Because $a(t) \in C([0, T])$, there exists $M > 0$ such that $\|a\| \leq M$ for any $t \in [0, T]$. Based on this consideration, we have

$$\|(\mathcal{R}_a y^f_0)(t)\| = \left\| \frac{y^f_0}{\Gamma(1-\alpha)} \int_0^t (t-\tau)^{-\alpha} a(\tau) d\tau \right\| \leq \frac{y^f_0 M t^{1-\alpha}}{\Gamma(2-\alpha)}. \tag{32}$$

Furthermore, suppose that the following relationship

$$\|(\mathcal{R}^i_a y^f_0)(t)\| \leq \frac{y^f_0 M^i t^{i(1-\alpha)}}{\Gamma(i(1-\alpha)+1)} \tag{33}$$

holds for any fixed $i \in \mathbb{N}$. Let us prove that the relationship (33) is also valid for $i+1$. According to the induction hypothesis, we get

$$\|(\mathcal{R}^{i+1}_a y^f_0)(t)\| = \frac{1}{\Gamma(1-\alpha)} \left\| \int_0^t (t-\tau)^{-\alpha} a(\tau)(\mathcal{R}^i_a y^f_0)(\tau) d\tau \right\| \tag{34}$$

$$\leq \frac{y^f_0 M^{i+1}}{\Gamma(1-\alpha)\Gamma(i(1-\alpha)+1)} \int_0^t (t-\tau)^{-\alpha} \tau^{i(1-\alpha)} d\tau. \tag{35}$$

Making use of a variable substitution $\tau = \omega t$, we have

$$\int_0^t (t-\tau)^{-\alpha} \tau^{i(1-\alpha)} d\tau = t^{(i+1)(1-\alpha)} \int_0^1 (1-\omega)^{-\alpha} \omega^{i(1-\alpha)} d\omega = t^{(i+1)(1-\alpha)} \frac{\Gamma(1-\alpha)\Gamma(i(1-\alpha)+1)}{\Gamma((i+1)(1-\alpha)+1)}, \tag{36}$$

where $B(\cdot, \cdot)$ is the Beta function defined as

$$B(z, w) = \int_0^1 (1-\tau)^{z-1} \tau^{w-1} d\tau, z, w > 0. \tag{37}$$

Here we used the relationship between the Beta function and the Gamma function:

$$B(z, w) = \frac{\Gamma(z)\Gamma(w)}{\Gamma(z+w)}. \tag{38}$$

So it has $\|(\mathcal{R}^{i+1}_a y^f_0)(t)\| \leq \frac{y^f_0 M^{i+1} t^{(i+1)(1-\alpha)}}{\Gamma((i+1)(1-\alpha)+1)}$. Hence, for any $i \in \mathbb{N}$, we have

$$\|(\mathcal{R}^i_a y^f_0)(t)\| \leq \frac{y^f_0 M^i t^{i(1-\alpha)}}{\Gamma(i(1-\alpha)+1)}. \tag{39}$$

That is to say, the series $\sum\limits_{i=0}^{\infty} (\mathcal{R}^i_a y^f_0)(t)$ is uniformly convergent with respect to $t \in [0, T]$, and the sum function is the unique solution of Equation (21). This completes the proof of this lemma. $\square$

With respect to this lemma, we have the following remarks.

Remark 1. *In particular, if $\alpha = 0$, then we have*

$$(\mathcal{R}_a^i y_0^f)(t) = \frac{y_0^f \left(\int_0^t a(\tau)d\tau \right)^i}{i!}, \quad i = 1, 2, \ldots. \tag{40}$$

Obviously, (40) is valid for $i = 1$. Suppose that (40) holds for any fixed i. Let us verify that (40) also holds for $i + 1$. According to the induction hypothesis, we have

$$(\mathcal{R}_a^{i+1} y_0^f)(t) = y_0^f \int_0^t a(\tau) \frac{\left(\int_0^\tau a(s)ds \right)^i}{i!} d\tau = y_0^f \int_0^t \frac{\left(\int_0^\tau a(s)ds \right)^i}{i!} d\left(\int_0^\tau a(s)ds \right) = \frac{y_0^f \left(\int_0^t a(\tau)d\tau \right)^{i+1}}{(i+1)!}. \tag{41}$$

So, (40) holds for any positive integer. Therefore, the solution of the following initial value problem

$$\frac{dY^f(t)}{dt} = a(t)Y^f(t), \quad Y^f(0) = y_0^f \tag{42}$$

is given by

$$Y^f(t) = y_0^f \exp\left(\int_0^t a(\tau)d\tau \right). \tag{43}$$

This coincides with the classical result.

Remark 2. *In [29], the author gave the solution of Equation (21) as*

$$Y^f(t) = y_0^f E_{1-\alpha}\left((1-\alpha) \int_0^t (t-\tau)^{-\alpha} a(\tau)d\tau \right). \tag{44}$$

We think the representation of the solution of Equation (21) is wrong. For example, $\alpha = \frac{1}{2}$, and $a(t) = t^\beta$, it has

$$\int_0^t (t-\tau)^{-\alpha} \tau^\beta d\tau = \frac{\Gamma(1-\alpha)\Gamma(1+\beta)}{\Gamma(2-\alpha+\beta)} t^{1-\alpha+\beta}. \tag{45}$$

Therefore, according to the result in [29], the solution is

$$Y^f(t) = y_0^f E_{\frac{1}{2}}\left(\frac{\Gamma(\frac{3}{2})\Gamma(1+\beta)}{\Gamma(\frac{3}{2}+\beta)} t^{\frac{1}{2}+\beta} \right). \tag{46}$$

However, we find that $Y^f(t)$ defined in (46) is not the solution of Equation (21).
In fact, by using our obtained result in Lemma 3, the solution of Equation (21) is

$$Y^f(t) = \sum_{i=0}^\infty \mathcal{R}_a^i y_0^f = y_0^f E_{\frac{1}{2}, 1+2\beta, 2\beta}\left(t^{\frac{1}{2}+\beta} \right), t \in [0, T]. \tag{47}$$

Next, we consider the solution of Equation (23).

Lemma 4. *Let $\frac{1}{2} < H < 1$ and $\sigma \in C([0, T])$. Then the solution of Equation (23) is*

$$Y^s(t) = y_0^s \exp\left(-H \int_0^t \tau^{2H-1}\sigma^2(\tau)d\tau + \int_0^t \sigma(\tau)dB_H(\tau) \right). \tag{48}$$

Proof. Let

$$Y^s(t) = y_0^s \exp \left(\int_0^t p_1(\tau)d\tau + \int_0^t p_2(\tau)dB_H(\tau) \right) \tag{49}$$

be the solution of Equation (23). Then, it satisfies Equation (23); i.e.,

$$dY^s(t) = y_0^s \sigma(t) \exp \left(\int_0^t p_1(\tau)d\tau + \int_0^t p_2(\tau)dB_H(\tau) \right) dB_H(t). \tag{50}$$

On the other hand, applying fractional Itô formula to $Y^s(t)$ in (49), we have

$$dY^s(t) = x_0 \exp \left(\int_0^t p_1(\tau)d\tau + \int_0^t p_2(\tau)dB_H(\tau) \right) \left(p_1(t) + Ht^{2H-1}p_2^2(t) \right) dt + p_2(t)dB_H(t). \tag{51}$$

Subtracting (50) from (51), we have

$$p_2(t) = \sigma(t), \quad p_1(t) = -Ht^{2H-1}\sigma^2(t). \tag{52}$$

Therefore, the solution of Equation (23) is

$$Y^s(t) = y_0^s \exp \left(-H \int_0^t \tau^{2H-1}\sigma^2(\tau)d\tau + \int_0^t \sigma(\tau)dB_H(\tau) \right). \tag{53}$$

The proof of this lemma is completed. $\square$

At this stage, we can establish the following theorem.

Theorem 1. *Let $a, b, \sigma \in C([0,T])$, $0 < \alpha < 1$, and $\frac{1}{2} < H < 1$. Then, the solution of Equation (20) is given by*

$$Y(t) = \exp \left(\int_0^t (b(\tau) - H\tau^{2H-1}\sigma^2(\tau))d\tau + \int_0^t \sigma(\tau)dB_H(\tau) \right) \sum_{i=0}^{\infty} \mathcal{R}_a^i y_0, \tag{54}$$

where $\mathcal{R}_a$ is defined as (28), and $\mathcal{R}_a^i$ denotes the i-times composition operator of $\mathcal{R}_a$.

We denote

$$\Phi(t) = \exp \left(\int_0^t b(\tau)d\tau - H \int_0^t \tau^{2H-1}\sigma^2(\tau)d\tau + \int_0^t \sigma(\tau)dB_H(\tau) \right) \sum_{i=0}^{\infty} \mathcal{R}_a^i. \tag{55}$$

One knows that Φ is the fundamental solution of Equation (20). In the following, we will show that Φ is invertible on $[0, T]$ in an algebraic sense.

Theorem 2. *Let Φ be the fundamental solution of Equation (20). Then Φ is invertible on $[0, T]$, and its inverse is*

$$\Phi^{-1} = \exp \left(-\int_0^t (b(\tau) - H\tau^{2H-1}\sigma^2(\tau))d\tau - \int_0^t \sigma(\tau)dB_H(\tau) \right) \sum_{i=0}^{\infty} (-1)^i \mathcal{R}_a^i, \tag{56}$$

where $\mathcal{R}_a$ is the operator defined as (28).

Proof. From Theorem 1, one knows that the following equation

$$dZ(t) = -\frac{1}{\Gamma(2-\alpha)}a(t)Z(t)(dt)^{1-\alpha} - \big(b(t) - 2Ht^{2H-1}\sigma^2(t)\big)Z(t)dt - \sigma(t)Z(t)dB_H(t), Z(0) = z_0 \quad (57)$$

has a unique solution $Z(t) = z_0\Psi(t)$, where $\Psi(t)$ is the fundamental solution of Equation (57) given by

$$\Psi(t) = \exp\left(-\int_0^t \big(b(\tau) - H\tau^{2H-1}\sigma^2(\tau)\big)d\tau - \int_0^t \sigma(\tau)dB_H(\tau)\right)\sum_{i=0}^{\infty}(-1)^i\mathcal{R}_a^i, \quad (58)$$

and follows that

$$d\Psi(t) = -\frac{1}{\Gamma(2-\alpha)}a(t)\Psi(t)(dt)^{1-\alpha} - \big(b(t) - 2Ht^{2H-1}\sigma^2(t)\big)\Psi(t)dt - \sigma(t)\Psi(t)dB_H(t). \quad (59)$$

Additionally, since Φ satisfies that

$$d\Phi(t) = \frac{1}{\Gamma(2-\alpha)}a(t)\Phi(t)(dt)^{1-\alpha} + b(t)\Phi(t)dt + \sigma(t)\Phi(t)dB_H(t). \quad (60)$$

Then, by the product rule, we have

$$d(\Phi\Psi) = \Psi(d\Phi) + \Phi(d\Psi) + d\Phi d\Psi = 0. \quad (61)$$

This implies that $\Phi\Psi \equiv constant$ on $t \in [0, T]$. On the other hand, we note that $\Phi(0)\Psi(0) = 1$. Thus, $\Phi(t)\Psi(t) \equiv 1$ on $t \in [0, T]$. This implies that Φ is invertible on $[0, T]$, and its inverse is Ψ. The proof is completed. $\square$

3.2. Solution Representation for Linear Nonhomogeneous Case

In this subsection, we consider the solution of Equation (2). We use the variation of constants parameters to find a particular solution Y_p of Equation (2). For this purpose, we define a random function

$$Y_p(t) = \Phi(t)c(t), \quad (62)$$

where $c(t)$ is an unknown random function with $c(0) = y_0$. Let us assume that $Y_p(t)$ is a solution of Equation (2).

By the product rule to Y_p, we have

$$dY_p(t) = d\Phi(t)c(t) + \Phi(t)dc(t) + d\Phi(t)dc(t). \quad (63)$$

Additonally, since Φ is invertible, it has

$$dc(t) = \Phi^{-1}\big(dY_p(t) - d\Phi(t)c(t) - d\Phi(t)dc(t)\big). \quad (64)$$

Furthermore, since $Y_p(t)$ is the solution of Equation (2) and Φ is the solution of Equation (20), we have

$$dc(t) = \Phi^{-1}(t)p(t)(dt)^{1-\alpha} + \Phi^{-1}(t)q(t)dt + \Phi^{-1}(t)v(t)dB_H(t) - \Phi^{-1}d\Phi(t)dc(t). \quad (65)$$

Additionally, since

$$d\Phi(t)dc(t) = 2Ht^{2H-1}v(t)\sigma(t)dt. \quad (66)$$

Therefore, we have

$$dc(t) = \Phi^{-1}(t)p(t)(dt)^{1-\alpha} + \Phi^{-1}(t)\big(q(t) - 2Ht^{2H-1}v(t)\sigma(t)\big)dt + \Phi^{-1}(t)v(t)dB_H(t), \quad (67)$$

and

$$
\begin{aligned}
c(t) \;=\;& c(0) + \int_0^t \Phi^{-1}(\tau)p(\tau)(d\tau)^{1-\alpha} + \int_0^t \Phi^{-1}(\tau)\big(q(\tau) - 2H\tau^{2H-1}v(\tau)\sigma(\tau)\big)d\tau \\
& + \int_0^t \Phi^{-1}(\tau)v(\tau)dB_H(\tau).
\end{aligned}
\tag{68}
$$

Thus, the solution $Y(t)$ of Equation (2) is

$$
\begin{aligned}
Y(t) \;=\;& \Phi(t)y_0 + \int_0^t \Phi(t,\tau)p(\tau)(d\tau)^{1-\alpha} + \int_0^t \Phi(t,\tau)\big(q(\tau) - 2H\tau^{2H-1}v(\tau)\sigma(\tau)\big)d\tau \\
& + \int_0^t \Phi(t,\tau)v(\tau)dB_H(\tau),
\end{aligned}
\tag{69}
$$

where $\Phi(t,\tau) = \Phi(t)\Phi^{-1}(\tau)$, Φ and Φ^{-1} are defined as (55) and (56), respectively.

Based on the above analysis, we can establish the following theorem.

Theorem 3. *Let* $a, p, q, v, \sigma \in C[0,T]$, $0 < \alpha < 1$ *and* $\frac{1}{2} < H < 1$. *Then the solution of Equation* (2) *is given by*

$$
\begin{aligned}
Y(t) \;=\;& \Phi(t)y_0 + \int_0^t \Phi(t,\tau)p(\tau)(d\tau)^{1-\alpha} + \int_0^t \Phi(t,\tau)\big(q(\tau) - 2H\tau^{2H-1}v(\tau)\sigma(\tau)\big)d\tau \\
& + \int_0^t \Phi(t,\tau)v(\tau)dB_H(\tau),
\end{aligned}
\tag{70}
$$

where $\Phi(t,\tau) = \Phi(t)\Phi^{-1}(\tau)$, Φ *and* Φ^{-1} *are defined as* (55) *and* (56), *respectively.*

4. Applications

In this section, we demonstrate some applications of our obtained results.

Example 1. *In this example, we consider a a mathematical model that can simulate the prices of financial instruments (e.g., stocks).*

Let $(\Omega, \mathcal{F}, \mathcal{P})$ *be a probability space, where* Ω *is called a sample space,* F *is a set of all events and possible statements about the prices on the market, and* P *is the usual probability measure. The price of an asset* Z_t *in classical Black–Scholes model is assumed to follow Geometric Brownian motion given by*

$$
dZ_t = \left(\mu + \frac{1}{2}\sigma^2\right)Z_t dt + \sigma Z_t dW_t, \; Z_0 = z_0,
\tag{71}
$$

where W_t *is the standard Brownian motion with respect to* $\mathcal{P}$, $\sigma > 0$ *is the diffusion parameter, and* $\mu \in \mathbb{R}$ *is the drift.*

The classical Black–Scholes model was certainly a breakthrough in the option pricing apparatus, because in the financial market, one needs to consider the influence of maturity time and the strike price on the financial derivatives or other factors. For these reasons, the Black–Scholes model with subdiffusion term is assumed to follow fractional Brownian motion given by [41]

$$
\frac{dZ_t}{dt} + aD_t^\alpha Z_t = \left(\mu + \frac{1}{2}\sigma^2\right)Z_t + \sigma Z_t \frac{dB_H(t)}{dt}, \; Z_0 = z_0,
\tag{72}
$$

where B_H *is the fractional Brownian motion with respect to* $\mathcal{P}$, $H \in (\frac{1}{2}, 1)$, $a > 0$ *is the subdiffusion parameter,* $\sigma > 0$ *is the diffusion parameter, and* $\mu \in \mathbb{R}$ *is the drift.*

For obtaining the solution of Equation (72), we divide three steps to solve it:

Step 1: According to (19), Equation (72) is equivalent to the following integral equation:

$$\begin{cases} dZ_t = \frac{a}{\Gamma(2-\alpha)} Z_t (dt)^{1-\alpha} + \left(\mu + \frac{1}{2}\sigma^2\right) Z_t dt + \sigma Z_t dB_H(t), \\ Z_0 = z_0. \end{cases} \tag{73}$$

So, we only need to solve (73). Furthermore, according to (21), (22), and (23), Z_t can be expressed as $Z_t = Z_t^f Z_t^d Z_t^s$, where Z_t^f is the solution of the following equation:

$$dZ_t^f = \frac{a}{\Gamma(2-\alpha)} Z_t^f (dt)^{1-\alpha}, Z_0^f = z_0^f, \tag{74}$$

Z_t^d is the solution of the following equation:

$$dZ_t^d = \left(\mu + \frac{1}{2}\sigma^2\right) Z_t^d dt, Z_0^d = z_0^d, \tag{75}$$

and Z_t^s is the solution of the following equation:

$$dZ_t^s = \sigma Z_t^s dB_H(t), Z_0^s = z_0^s, \tag{76}$$

and also $z_0^f z_0^d z_0^s = z_0$.

Step 2: Solve Equations (74)–(76), respectively. According to Lemma 3, the solution of Equation (74) is $Z_t^f = z_0^f E_{1-\alpha}(at^{1-\alpha})$. According to Lemma 4, the solution of Equation (76) is $Z_t^s = z_0^s \exp\left(-\frac{\sigma^2 t^{2H}}{2} + \sigma B_H(t)\right)$.

Step 3: According to Theorem 1, Z_t is given by

$$Z_t = z_0 E_{1-\alpha}(at^{1-\alpha}) \exp\left(-\frac{\sigma^2 t^{2H}}{2} + \sigma B_H(t) + \left(\mu + \frac{1}{2}\sigma^2\right)t\right). \tag{77}$$

Example 2. *Consider the following fractional stochastic partial differential equation*

$$\frac{\partial U(x,t)}{\partial t} + D_t^\alpha U(x,t) = -k_{p_1}(-\triangle)^{\frac{p_1}{2}} U(x,t) - k_{p_2}(-\triangle)^{\frac{p_2}{2}} U(x,t) + U(x,t)\frac{dB_H(t)}{dt}, \tag{78}$$

with the nonhomogeneous Dirichlet boundary conditions

$$U(0,t) = U(L,t) = 0, \tag{79}$$

and the initial condition

$$U(x,0) = \phi(x), \tag{80}$$

where $(x,t) \in [0,L] \times [0,T]$ (L and T are constants), $0 < \alpha < 1, 0 < p_1 \leq 1, 1 < p_2 \leq 2, \frac{1}{2} < H < 1$, and $\varphi(x)$ is a random function.

According to Lemma 2, the eigenvalues λ_n^2 ($n = 1,2,\ldots$) of the operator $(-\triangle)$ with the homogeneous boundary conditions are $\lambda_n^2 = n^2\pi^2/L^2$, and the corresponding eigenfunctions are $\varphi_n(x) = \sin(n\pi x/L), n = 1,2,\ldots$. Then we set

$$U(x,t) = \sum_{n=1}^{\infty} U_n(t) \sin(n\pi x/L). \tag{81}$$

Substituting (81) into (78) and (80) leads to the following equation:

$$\frac{dU_n(t)}{dt} + D_t^\alpha U_n(t) = -k_{p_1}\lambda_n^{p_1} U_n(t) - k_{p_2}\lambda_n^{p_2} U_n(t) + U_n(t)\frac{dB_H(t)}{dt}, \tag{82}$$

with the initial condition

$$U_n(0) = \frac{2}{L}\int_0^L \phi(x)\sin(n\pi x/L)dx. \tag{83}$$

By Theorem 1, the solution of Equation (82) with the initial condition (83) is

$$U_n(t) = U_n(0)\exp\left(\left(-k_{p_1}\lambda_n^{p_1} - k_{p_2}\lambda_n^{p_2}\right)t - \frac{t^{2H}}{H} + B_H(t)\right)E_{1-\alpha}(t^{1-\alpha}). \tag{84}$$

Therefore, the solution of Equation (78) with the boundary conditions (79) and the initial condition (80) is

$$u(x,t) = \sum_{n=1}^\infty U_n(0)\exp\left(\left(-k_{p_1}\lambda_n^{p_1} - k_{p_2}\lambda_n^{p_2}\right)t - \frac{t^{2H}}{H} + B_H(t)\right)E_{1-\alpha}(t^{1-\alpha})\sin(n\pi x/L). \tag{85}$$

Example 3. *Consider the following fractional stochastic partial differential equation*

$$\frac{\partial U(x,t)}{\partial t} + D_t^\alpha U(x,t) = -k_{p_1}(-\triangle)^{\frac{p_1}{2}}U(x,t) - k_{p_2}(-\triangle)^{\frac{p_2}{2}}U(x,t) + v(t)\frac{dB_H(t)}{dt} + f(x,t), \tag{86}$$

with the nonhomogeneous Dirichlet boundary conditions

$$U(0,t) = U(L,t) = 0, \tag{87}$$

and the initial condition

$$U(x,0) = \psi(x), \tag{88}$$

where $(x,t) \in [0,L] \times [0,T]$ *(L and T are constants),* $0 < \alpha < 1, 0 < p_1 \le 1, 1 < p_2 \le 2, \frac{1}{2} < H < 1,$ *and* $\varphi(x)$ *is a random function.*

According to Lemma 2, the eigenvalues λ_n^2 $(n = 1,2,\ldots)$ of the operator $(-\triangle)$ with the homogeneous boundary conditions are $\lambda_n^2 = n^2\pi^2/L^2$, and the corresponding eigenfunctions are $\varphi_n(x) = \sin(n\pi x/L), n = 1,2,\ldots$. Then we set

$$U(x,t) = \sum_{n=1}^\infty U_n(t)\sin(n\pi x/L), \quad f(x,t) = \sum_{n=1}^\infty f_n(t)\sin(n\pi x/L). \tag{89}$$

Substituting (89) into (86) and (88) leads to the following equation

$$\frac{dU_n(t)}{dt} + D_t^\alpha U_n(t) = -k_{p_1}\lambda_n^{p_1} U_n(t) - k_{p_2}\lambda_n^{p_2} U_n(t) + v(t)\frac{dB_H(t)}{dt} + f_n(t), \tag{90}$$

with the initial condition

$$U_n(0) = \frac{2}{L}\int_0^L \psi(x)\sin(n\pi x/L)dx. \tag{91}$$

By Theorem 3, the solution of Equation (86) with the initial condition (87) is

$$U_n(t) = \Phi(t)U_n(0) + \int_0^t \Phi(t,\tau)f_n(\tau)(d\tau)^{1-\alpha} + \int_0^t \Phi(t,\tau)f_n(\tau)d\tau + \int_0^t \Phi(t,\tau)v(\tau)dB_H(\tau), \quad (92)$$

where

$$\Phi(t) = \exp\left(-(k_{p_1}\lambda_n^{p_1} + k_{p_2}\lambda_n^{p_2})t\right)E_{1-\alpha}(t^{1-\alpha}), \quad (93)$$

$$\Phi^{-1}(t) = \exp\left((k_{p_1}\lambda_n^{p_1} + k_{p_2}\lambda_n^{p_2})t\right)E_{1-\alpha}(-t^{1-\alpha}), \quad (94)$$

and $\Phi(t,\tau) = \Phi(t)\Phi^{-1}(\tau)$. Therefore, the solution of Equation (86) with the boundary condition (87) and the initial condition (88) is

$$U(x,t) = \sum_{n=1}^{\infty} U_n(t)\sin(n\pi x/L), \quad (95)$$

where $U_n(t)$ is defined in (92).

5. Conclusions

In this paper, we gave analytical solutions of multi-time scale fractional stochastic differential equations driven by fractional Brownian motions. We first decomposed the homogeneous multi-time scale fractional stochastic differential equation driven by fractional Brownian motion into independent differential subequations, and gave its analytical solution. Then, we used the variation of constants parameters to obtain the solution of the nonhomogeneous multi-time scale fractional stochastic differential equation driven by fractional Brownian motion. Finally, we demonstrated the applicability of our obtained results in solving FSDEs.

FSPDEs are an important class of differential equations. In this paper, we combined our obtained results about fractional stochastic ordinary differential equations and spectral representation technique to give the analytical solutions of some FSPDEs. In the future, we will investigate entropy analyses including permutation entropy, fractional permutation entropy, sample entropy, and fractional sample entropy with the help of our obtained analytical solutions in some practical problems. On the other hand, we plan to use the obtained analytical solutions of FSPDEs to assess the computational performance and accuracy of their numerical solutions which we will develop.

Acknowledgments: The work of the Xiao-Li Ding was supported by the Natural Science Foundation of China (11501436) and Young Talent fund of University Association for Science and Technology in Shaanxi, China (20170701). The work of Juan J. Nieto has been partially supported by the AEI of Spain under Grant MTM2016-75140-P and co-financed by European Community fund FEDER, and XUNTA de Galicia under grants GRC2015-004 and R2016/022.

Author Contributions: Xiao-Li Ding established the main theorem and wrote the main text. Juan J. Nieto examined the fractional model, and reviewed the text. All the authors have read and approved the final manuscript.

Conflicts of Interest: The authors declare no conflict of interest.

References

1. Benson, D.A.; Wheatcraft, S.W.; Meerschaert, M.M. Application of a fractional advection-dispersion equation. *Water Resour. Res.* **2000**, *36*, 1403–1412.
2. Hilfer, R. *Applications of Fractional Calculus in Physics*; World Scientific: Singapore, 2000.
3. Vázquez, J.L. The mathematical theories of diffusion: Nonlinear and fractional diffusion. In *Lecture Notes in Mathematics*; Morel, J.-M., Cachan, ENS, Teissier, B., Paris 7, U., Eds; Springer: Berlin/Heidelberg, Germany, 2017, pp. 205–278.

4. Hall, M.G.; Barrick, T.R. From diffusion-weighted MRI to anomalous diffusion imaging. *Magn. Reson. Med.* **2008**, *59*, 447–455.

5. Cesbron, L.; Mellet, A.; Trivisa, K. Anomalous transport of particles in plasma physics. *Appl. Math. Lett.* **2012**, *25*, 2344–2348.

6. Metzler, R.; Klafter, J. The random walk's guide to anomalous diffusion: A fractional dynamics approach. *Phys. Rep.* **2000**, *339*, 1–77.

7. Postnikov, E.B.; Sokolov, I.M. Model of lateral diffusion in ultrathin layered films. *Phys. A* **2012**, *391*, 5095–5101.

8. Wang, Y.; Zheng, S.; Zhang, W.; Wang, J. Complex and Entropy of Fluctuations of Agent-Based Interacting Financial Dynamics with Random Jump. *Entropy* **2017**, *19*, 512, doi:10.3390/e19100512.

9. San-Millan, A.; Feliu-Talegon, D.; Feliu-Batlle, V.; Rivas-Perez, R. On the Modelling and Control of a Laboratory Prototype of a Hydraulic Canal Based on a TITO Fractional-Order Model. *Entropy* **2017**, *19*, 401, doi:10.3390/e19080401.

10. Alsaedi, A.; Nieto, J.J.; Venktesh, V. Fractional electrical circuits. *Adv. Mech. Eng.* **2015**, *7*, 1–7.

11. Ladde, G.S.; Wu, L. Development of nonlinear stochastic models by using stock price data and basic statistics. *Neutral Parallel Sci. Comput.* **2010**, *18*, 269–282.

12. Tien, D.N. Fractional stochastic differential equations with applications to finance. *J. Math. Anal. Appl.* **2013**, *397*, 334–348.

13. Farhadi, A.; Erjaee, G.H.; Salehi, M. Derivation of a new Merton's optimal problem presented by fractional stochastic stock price and its applications. *Comput. Math. Appl.* **2017**, *73*, 2066–2075.

14. Arnold, L. *Stochastic Differential Equations: Theory and Applications*; Wiley: New York, NY, USA, 1974.

15. Mandelbrot, B.; Van Ness, J. Fractional Brownian motions, fractional noises and applications. *SIAM Rev.* **1968**, *10*, 422–437.

16. Liu, J.; Yan, L. Solving a nonlinear fractional stochastic partial differential equation with fractional noise. *J. Theor. Probab.* **2016**, *29*, 307–347.

17. Xia, D.; Yan, L. Some properties of the solution to fractional heat equation with a fractional Brownian noise. *Adv. Differ. Equ.* **2017**, *2017*, 107, doi:10.1186/s13662-017-1151-0.

18. Taheri, Z.; Javadi, S.; Babolian, E. Numerical solution of stochastic fractional integro-differential equation by the spectral collocation method. *J. Comput. Appl. Math.* **2017**, *321*, 336–347.

19. Tamilalagan, P.; Balasubramaniam, P. Moment stability via resolvent operators of fractional stochastic differential inclusions driven by fractional Brownian motion. *Appl. Math. Comput.* **2017**, *305*, 299–307.

20. Asogwa, S.A.; Nane, E. Intermittency fronts for space-time fractional stochastic partial differential equations in $(d+1)$ dimensions. *Stoch. Process. Appl.* **2017**, *127*, 1354–1374.

21. Li, X.; Yang, X. Error Estimates of Finite Element Methods for Stochastic Fractional Differential Equations. *J. Comput. Math.* **2017**, *35*, 346–362.

22. Kubilius, K.; Skorniakov, V.; Ralchenko, K. The rate of convergence of the Hurst index estimate for a stochastic differential equation. *Nonlinear Anal. Model. Control* **2017**, *22*, 273–284.

23. Hernández-Hernández, M.E.; Kolokoltsov, V.N. On the solution of two-sided fractional ordinary differential equations of Caputo type. *Fract. Calc. Appl. Anal.* **2016**, *19*, 1393–1413.

24. Nane, E.; Ni, Y. Stochastic solution of fractional Fokker–Planck equations with space–time-dependent coefficients. *J. Math. Anal. Appl.* **2016**, *442*, 103–116.

25. Kloeden, P. E.; Platen, E. *Numerical Solution of Stochastic Differential Equations*; Springer: Berlin/Heidelberg, Germany, 1992.

26. Jiang, Y.; Wei, T.; Zhou, X. Stochastic generalized Burgers equations driven by fractional noises. *J. Differ. Equ.* **2011**, *252*, 1934–1961.

27. Hu, Y. Heat equations with fractional white noise potentials. *Appl. Math. Optim.* **2001**, *43*, 221–243.

28. Alós, E.; Nualart, D. Stochastic integration with respect to the fractional Brownian motion. *Stoch. Stoch. Rep.* **2003**, *75*, 129–152.

29. Pedjeu, J.-C.; Ladde, G.S. Stochastic fractional differential equations: Modeling, method and analysis. *Chaos Solitons Fractals* **2012**, *45*, 279–293.

30. Singh, J.; Kumar, D.; Qurashi, M.A.; Baleanu, D. A Novel Numerical Approach for a Nonlinear Fractional Dynamical Model of Interpersonal and Romantic Relationships. *Entropy* **2017**, *19*, 375, doi:10.3390/e19070375.

31. Kamrani, M.; Jamshidi, N. Implicit Euler approximation of stochastic evolution equations with fractional Brownian motion. *Commun. Nonlinear Sci. Numer. Simulat.* **2017**, *44*, 1–10.

32. Wang, X.; Qi, R.; Jiang, F. Sharp mean-square regularity results for SPDEs with fractional noise and optimal convergence rates for the numerical approximations. *BIT Numer. Math.* **2016**, *57*, 557–585.

33. Jiang, H.; Liu, F.; Turner, I.; Burrage, K. Analytical solutions for the multi-term time–space Caputo–Riesz fractional advection–diffusion equations on a finite domain. *J. Math. Anal. Appl.* **2012**, *389*, 1117–1127.

34. Chang, S.-Y.A.; González, M.D.M. Fractional Laplacian in conformal geometry. *Adv. Math.* **2011**, *226*, 1410–1432.

35. Ding, X.L.; Jiang, Y.L. Analytical solutions for the multi-term time–space fractional advection–diffusion equations with mixed boundary conditions. *Nonlinear Anal. Real World Appl.* **2013**, *14*, 1026–1033.

36. Ding, X.L.; Nieto, J.J. Analytical solutions for coupling fractional partial differential equations with Dirichlet boundary conditions. *Commun. Nonlinear Sci. Numer. Simulat.* **2017**, *52*, 165–176, doi:10.1016/j.cnsns.2017.04.020.

37. Kilbas, A.A.; Srivastava, H.M.; Trujillo, J.J. *Theory and Applications of Fractional Differential Equations*; North-Holland Mathematics Studies; Elsevier Science B.V.: Amsterdam, The Netherlands, 2006; Volume 204.

38. Itô, K. *Stochastic Differential Equations*; Wiley Interscience: New York, NY, USA, 1978.

39. Bender, C. An Itô formula for generalized functionals of a fractional Brownian motion with arbitrary Hurst parameter. *Stoch. Process. Appl.* **2003**, *104*, 81–106.

40. Nualart, D. The Malliavin Calculus and Related Topics, Springer-Verlag: Berlin/Heidelberg, Germany, 2006.

41. Karipova, G.; Magdziarz, M. Pricing of basket options in subdiffusive fractional Black-Scholes model. *Chaos Solitons Fractals* **2017**, *102*, 245–253.

MDPI
St. Alban-Anlage 66
4052 Basel
Switzerland
Tel. +41 61 683 77 34
Fax +41 61 302 89 18
www.mdpi.com

Entropy Editorial Office
E-mail: entropy@mdpi.com
www.mdpi.com/journal/entropy